Shortcut
Algebra I

Andrew Marx

KAPLAN

PUBLISHING

New York • Chicago

This publication is designed to provide accurate and authoritative information in regard to the subject matter covered. It is sold with the understanding that the publisher is not engaged in rendering legal, accounting, or other professional service. If legal advice or other expert assistance is required, the services of a competent professional should be sought.

Editorial Director: Jennifer Farthing
Editor: Ruth Baygell
Production Artist: Baldur Gudbjornsson
Cover Designer: Carly Schnur

Published by Kaplan Publishing, a division of Kaplan, Inc.
888 Seventh Ave.
New York, NY 10106

Printed in the United States of America

June 2006

10 9 8 7 6 5 4 3 2 1

ISBN-13: 978-1-4195-4166-7
ISBN-10:   1-4195-4166-8

Kaplan Publishing books are available at special quantity discounts to use for sales promotions, employee premiums, or educational purposes. Please call our Special Sales Department to order or for more information at 800-621-9621, ext. 4444, e-mail kaplanpubsales@kaplan.com, or write to Kaplan Publishing, 30 South Wacker Drive, Suite 2500, Chicago, IL 60606-7481.

TABLE OF CONTENTS

Introduction vii

Diagnostic Quiz 1
Answer Explanations 7
Question Correlation Chart 24

Chapter 1: Number Properties 25
What Are Number Properties?. 25
Concepts to Help You 25
Steps You Need to Remember 29
Step-by-Step Illustration of the 5 Most Common Question Types. 36
Chapter Quiz 44
Answer Explanations 46

Chapter 2: Algebraic Expressions 51
What Are Algebraic Expressions? 51
Concepts to Help You 51
Steps You Need to Remember 55
Step-by-Step Illustration of the 5 Most Common Question Types. 56
Chapter Quiz 63
Answer Explanations 64

Chapter 3: Factors 69
What Are Factors? 69
Concepts to Help You 69
Steps You Need to Remember 75
Step-by-Step Illustration of the 5 Most Common Question Types. 77
Chapter Quiz 81
Answer Explanations 83

Chapter 4: Roots, Radicals and Powers..........................87

What Are Powers, Roots, and Radicals 87

Concepts to Help You 88

Steps You Need to Remember 95

Step-by-Step Illustration of the 5 Most Common Question Types.......96

Chapter Quiz 101

Answer Explanations 102

Chapter 5: Equation Solving 105

What Is Equation Solving?......................................105

Concepts to Help You..105

Steps You Need to Remember 109

Step-by-Step Illustration of the 5 Most Common Question Types......111

Chapter Quiz 117

Answer Explanations 118

Chapter 6: Coordinate Geometry..............................123

What Is Coordinate Geometry? 123

Concepts to Help You..123

Steps You Need to Remember 131

Step-by-Step Illustration of the 5 Most Common Question Types......132

Chapter Quiz 138

Answer Explanations 140

Chapter 7: Graphing Linear Equations 145

What Is Linear Equation Graphing?..............................145

Concepts to Help You..146

Steps You Need to Remember 148

Step-by-Step Illustration of the 5 Most Common Question Types......150

Chapter Quiz 156

Answer Explanations 159

Chapter 8: Systems of Equations 163
What Are Systems of Equations? 163
Concepts to Help You 163
Steps You Need to Remember 165
Step-by-Step Illustration of the 5 Most Common Question Types. 166
Chapter Quiz 172
Answer Explanations 174

Chapter 9: Linear and Compound Inequalities 179
What Are Inequalities? 179
Concepts to Help You 179
Steps You Need to Remember 184
Step-by-Step Illustration of the 5 Most Common Question Types. 186
Chapter Quiz 193
Answer Explanations 198

Chapter 10: Polynomials 203
What Are Polynomials? 203
Concepts to Help You 204
Steps You Need to Remember 208
Step-by-Step Illustration of the 5 Most Common Question Types. 209
Chapter Quiz 214
Answer Explanations 215

Chapter 11: Quadratic Equations 219
What Are Quadratic Equations? 219
Concepts to Help You 219
Steps You Need to Remember 223
Step-By-Step Illustration of the 5 Most Common Question Types 225
Chapter Quiz 230
Answer Explanations 231

Chapter 12: Algebraic Fractions 235
What Are Algebraic Fractions? 235
Concepts to Help You 235
Steps You Need to Remember 238
Step-by-Step Illustration of the 5 Most Common Question Types 239
Chapter Quiz 245
Answer Explanations 247

Chapter 13: Word Problems 251
What are Word Problems? 251
Concepts to Help You 251
Steps You Need to Remember 254
Step-by-Step Illustration of the 5 Most Common Question Types 255
Chapter Quiz 260
Answer Explanations 261

INTRODUCTION

If you have picked up this book, you might be one of the many people who thinks Algebra I is an intimidating topic, but who needs to have basic knowledge of the subject. Maybe you're a first-time algebra student who needs some additional guidance in a high school or college class. Maybe you're a student gearing up for a standardized test in math. Maybe you're a professional in a business or health field, and you want to learn the basics as they apply to your work.

If you are one of these people, this book is designed just for you! *Shortcut Algebra I* offers an easy-to-understand approach that will guide you through the maze of problems and proofs that comprise basic Algebra. With over 200 step-by-step examples and practice questions, you'll be well on your way to feeling confident and at-ease with this challenging subject.

Algebra makes many students nervous because questions can take so many different forms. There are many ways for teachers and test question writers to vary the same basic kind of problem. Take a look at these:

- What is the solution to $6 \cdot x = 24$?
- If $6 \cdot x = 24$, $x = ?$
- Six times what number is 24?
- If a rectangle has a length of 6 and an area of 24, what is the width?

All of the above are just different forms of the same basic problem. Identifying the basic problem in a way that allows you to set things up for the solution is half the battle. Algebra I involves a small number of concepts and skills tested in different ways. Once you realize that, you'll appreciate the importance of setting up problems so that solving them becomes straightforward. Once a problem is set up properly, it's just a matter of going through standard steps to get to the solution.

Shortcut Algebra I begins with a review of common Algebra I topics that are the foundation for solving equations, which is mainly what Algebra is all about. Next comes a step-by-step introduction to the heart of the subject. Throughout, our discussion includes examples of how to apply concepts to concrete, real-world problems—essential for bringing Algebra I out of the realm of the abstract and into everyday light.

To use the book to the fullest advantage, start by taking the Diagnostic Quiz. Following the quiz, a Correlation Chart guiding you to the appropriate chapter for each question. Detailed answer explanations will help you identify the areas in which you need work.

If you have enough time, we recommend that you work through the entire book, because each chapter builds on information presented in previous chapters. For example, to understand quadratic equations, you need to understand polynomials, and to understand polynomials, you need to understand powers. Similarly, working with polynomials requires that you first understand how to simplify algebraic expressions.

Each core chapter is structured to identify key concepts and outline steps that will help you solve the most common types of questions. You'll learn how to apply those steps to real problems. As you work your way to a solution, detailed explanations will walk you through problem-solving techniques and useful strategies. Each chapter concludes with a brief quiz, which will help you to evaluate yourself and apply your understanding.

By the time you reach the end of *Shortcut Algebra I*, we're confident that you will see Algebra I in a whole new light—and you'll be amazed that it took so little time to get from where you started to a clear understanding of the basics. You'll be well on your way to mastering the essential skills and concepts that you need to succeed.

Good luck—and enjoy the shortcut!

Diagnostic Test

The following brief exam includes questions on every major topic in algebra I. It will help you to identify the areas where you may already be up to speed, as well as the areas you'll need to carefully review.

After you have scored your test, you can analyze your results with the help of the chart included at the end; it matches every question with a section of the book. You can use the chart to jump ahead if you need a quick review.

Keep in mind that this book covers a great deal more than any single test can. This Diagnostic, however, can still give you a good idea of how much attention you'll have to give to each chapter.

1. Evaluate $4 - 5 \times 3 - 2$.
   - (A) -21
   - (B) -13
   - (C) -9
   - (D) -5
   - (E) -1

2. What is the sum of 11.25, 0.054, and $\frac{3}{8}$?
   - (A) 11.454
   - (B) 11.679
   - (C) 11.734
   - (D) 11.930
   - (E) 12.165

3. What is the multiplicative inverse of 0.8?
   - (A) -0.8
   - (B) 0.2
   - (C) 1.25
   - (D) 1.2
   - (E) 8

4. Which of these is a simplified form of $4(3s + 4t) - 2t(2s + 5)$?
   - (A) $7s - 4ts - 2t$
   - (B) $10s - 2ts + 2t$
   - (C) $10s + 2ts - 10t$
   - (D) $12s - 4ts - 10t$
   - (E) $12s - 4ts + 6t$

5. The associative property shows that $8 + (4e + 9) =$

   (A) $(8 + 4e) + 9$

   (B) $(4e + 9) + 8$

   (C) $4e + 17$

   (D) $21e$

   (E) $(8 + 4e) + (8 + 9)$

6. If $a = 7$ and $b = 4$, then $3a + 5 - 2b =$

   (A) 12

   (B) 18

   (C) 28

   (D) 34

   (E) 38

7. What is the least common multiple of 6 and $8x$?

   (A) 2

   (B) $14x$

   (C) $24x$

   (D) $48x$

   (E) 48

8. $\dfrac{2x}{9} + \dfrac{3x}{12} =$

   (A) $\dfrac{5x}{21}$

   (B) $\dfrac{x}{2}$

   (C) $\dfrac{17x}{36}$

   (D) $\dfrac{13x}{24}$

   (E) $\dfrac{5x}{9}$

9. $\sqrt{50x^2} =$

   (A) $5x\sqrt{2x}$

   (B) $5\sqrt{2x}$

   (C) $x^2\sqrt{50}$

   (D) $5x\sqrt{2}$

   (E) $50x$

10. $\sqrt[3]{-64x^{27}} =$

    (A) $-64x^3$

    (B) $-24x^3$

    (C) $-12x^3$

    (D) $-8x^9$

    (E) $-4x^9$

11. If $4x + 7 = 51$, $x =$

   (A) 7

   (B) 9

   (C) 11

   (D) 13

   (E) 15

12. What is the solution of $52 - 3y = 16$?

   (A) −13

   (B) −12

   (C) 11

   (D) 12

   (E) 13

13. What are the coordinates of point *A* on the coordinate grid?

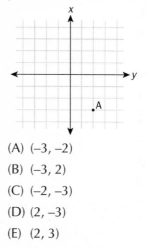

   (A) (−3, −2)

   (B) (−3, 2)

   (C) (−2, −3)

   (D) (2, −3)

   (E) (2, 3)

14. What are the coordinates of the point four units to the left and eight units above the point (3, −6) on the coordinate grid?

   (A) (−1, 2)

   (B) (7, 2)

   (C) (7, −14)

   (D) (11, −10)

   (E) (11, −2)

15. What is the slope of the graph of $y = \dfrac{x}{4} - 5$?

   (A) −5

   (B) $-\dfrac{5}{4}$

   (C) $\dfrac{1}{4}$

   (D) $\dfrac{4}{5}$

   (E) 4

16. Which of the following is the graph of $y = 3x - 1$?

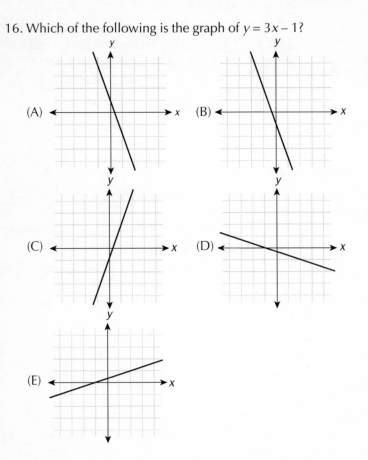

17. If $6x + 7y = 53$ and $8x - 2y = 14$, what is the value of x?

   (A) 2

   (B) 3

   (C) 4

   (D) 5

   (E) 6

18. If $x - y = 17$ and $x + y = 63$, $y =$

   (A) 20

   (B) 23

   (C) 34

   (D) 40

   (E) 46

19. If $2x - 6 > 8$, then
    (A) $x > -5$
    (B) $x > -2$
    (C) $x > 1$
    (D) $x > 7$
    (E) $x > 10$

20. If $4x - 9y \geq -12$, then
    (A) $y \leq \dfrac{4x + 12}{9}$
    (B) $y \geq \dfrac{4x}{9} + \dfrac{4}{3}$
    (C) $y \geq \dfrac{4x - 12}{9}$
    (D) $y \leq \dfrac{4x}{9} - \dfrac{4}{3}$
    (E) $y \geq \dfrac{4x - 4}{3}$

21. $(4x^2 + 2x)(3x + 2) =$
    (A) $12x^3 + 4x$
    (B) $12x^3 + 4x^2$
    (C) $12x^3 + 6x^2$
    (D) $12x^3 + 6x^2 + 4x$
    (E) $12x^3 + 14x^2 + 4x$

22. $(4x^4 + 3x^2 + 3) + (5x^3 + 4x^2 + 2x)$
    (A) $4x^4 + 5x^3 + 7x^2 + 5x$
    (B) $4x^4 + 5x^3 + 7x^2 + 2x + 3$
    (C) $4x^4 + 8x^3 + 4x^2 + 2x + 3$
    (D) $9x^4 + 7x^2 + 5x$
    (E) $9x7 + 7x4 + 5x$

23. Which of the following is a factor of $x^2 - 7x + 10$?
    (A) $x - 7$
    (B) $x - 5$
    (C) $x + 1$
    (D) $x + 2$
    (E) $x + 10$

24. What are the solutions of $x^2 + 6x - 16 = 0$?
    (A) -2 and 8
    (B) -6 and 16
    (C) 2 and -8
    (D) 6 and -8
    (E) 6 and -16

25. If $2x^2 + 6x + 3 = 0$
    (A) $-6 \pm 2\sqrt{3}$
    (B) $-6 \pm \dfrac{\sqrt{3}}{2}$
    (C) $\dfrac{-3 \pm \sqrt{3}}{2}$
    (D) $\dfrac{-3 \pm 2\sqrt{3}}{2}$
    (E) $-3 \pm \dfrac{\sqrt{3}}{2}$

26. $\dfrac{3x-6}{4x+12}$ is undefined if $x =$

    (A) -3

    (B) -2

    (C) 2

    (D) 3

    (E) 4

27. $\dfrac{24x^6y^3z^8}{6x^2yz^4} =$

    (A) $4x^3y^3z^2$

    (B) $4x^3y^2z^4$

    (C) $4x^4y^2z^4$

    (D) $4x^5y^2z^7$

    (E) $4x^8y^5z^{12}$

28. Water is running from a faucet at a rate of 3 fluid ounces per second. At this rate the faucet will fill a pot with 48 fluid ounces of water in how many seconds?

    (A) 16

    (B) 18

    (C) 24

    (D) 45

    (E) 51

29. After having lunch in a restaurant, Darren paid $12.48, which included the price of the meal, a 15% tip, and $0.52 for sales tax. What was the price of the meal?

    (A) $10.17

    (B) $10.33

    (C) $10.40

    (D) $11.30

    (E) $11.81

30. A rectangular room has a perimeter of 52 feet and an area of 168 square feet. What are the dimensions of the room?

    (A) 7 ft. by 24 ft.

    (B) 8 ft. by 21 ft.

    (C) 10 ft. by 16 ft.

    (D) 11 ft. by 15 ft.

    (E) 12 ft. by 14 ft.

ANSWERS

1. B	9. D	17. B	25. C
2. B	10. E	18. B	26. A
3. C	11. C	19. D	27. C
4. E	12. D	20. A	28. A
5. A	13. D	21. E	29. C
6. B	14. A	22. B	30. E
7. C	15. C	23. B	
8. C	16. C	24. C	

EXPLANATIONS

1. B

Evaluating or simplifying an arithmetic expression like this is a matter of performing the operations (multiplication, subtraction) *in the correct order*. When there are no parentheses or exponents, you always perform multiplication (or division) *before* addition or subtraction. Start at the left of the expression and look for the first pair of numbers connected by a multiplication or division sign:

$4 - \underline{5 \times 3} - 2$

The 5 and 3 are connected by the multiplication sign, so you need to start with that operation. You could group them with parentheses, and then multiply:

$4 - (5 \times 3) - 2 = 4 - 15 - 2$

So now there is no other operation left here but subtraction. Start again at the left, and perform one operation at a time. Since $4 - 15 = -11$,

$4 - 15 - 2 = -11 - 2$

Now subtract 2 from -11:

$-11 - 2 = -13$

So (B) is the answer. To get this answer, you would need to know how to do subtraction involving negative numbers, as well the rules for carrying out operations in the right order. If you do not follow the correct order of operations and simply perform them from left to right (subtracting 5 from 4 before multiplying 5 by 3), you might come to the incorrect (D), –5. If you performed the operations from right to left rather than left to right, you would get –1, choice (E)

2. B

This question involves several of the number properties covered in Chapter 1. Algebra problems often require you to add decimals. You will also need to be able convert fractions to decimals.

Before you can add these three numbers, you'll have to convert the fraction to a decimal. You can always do this by dividing the numerator by the denominator. This is very easy if you have a calculator, but it can be done by hand. It would help to remember that $\frac{1}{8} = 0.125$. Some fractions appear often on tests, and you can work faster by memorizing their decimal values. Once you have the value of 0.125, you can multiply that by 3 to get 0.375, the decimal value of $\frac{3}{8}$.

Now you can add the three decimals: 11.25, 0.054, and 0.375. Arrange them so the decimal points line up, and then add the numbers column by column, carrying over when necessary. It is important to line up the decimal points, rather than lining up numbers by the first or last digit.

You can attach zeroes to the end of a decimal, so that each number goes to the same decimal place. We'll do that with 11.25

$$
\begin{array}{r}
11.250 \\
0.054 \\
+\ 0.375 \\
\hline
11.679
\end{array}
$$

So Choice (B) is the correct answer. It is easy to come to the wrong answer when you line up the decimals incorrectly. You would get 12.165, Choice (E), by adding 0.54 instead of 0.054. One might also try to add the decimals 0.25, 0.054 and 0.375 by ignoring the decimal place and adding the numbers as though they were integers. Merely adding 25, 54, and 375 to get 454 might lead you to mistakenly pick 11.454, Choice (A).

3. C

The multiplicative inverse is a basic number property discussed in chapter 1. A number and its multiplicative inverse have a product of 1. Reciprocal is another name for multiplicative inverse. So you're really being asked to find the number that gives you 1 when multiplied by 0.8:

$0.8 \times ? = 1$

It might help to rewrite the decimal 0.8 as a fraction: $0.8 = \frac{4}{5}$

So you need find the unknown in $\frac{4}{5} \times ? = 1$.

To find the multiplicative inverse of a fraction, divide the denominator by the numerator:

$4\overline{)5} = 1.25$

You can switch the numerator (the top number) and the denominator (the bottom number). So the inverse of $\frac{4}{5}$ is $\frac{5}{4}$. Check this by multiplying the numbers:

$\frac{4}{5} \times \frac{5}{4} = \frac{20}{20}$, and all fractions that have the same number on top and bottom equal 1.

So we know that $\frac{5}{4}$ is the multiplicative inverse of $\frac{4}{5}$. We just have to convert it to a decimal.

Note that $\frac{5}{4} = 5 \times \frac{1}{4}$. $\frac{1}{4} = 0.25$, so 5 x 0.25 = 1.25, so (C) is correct.

(A) might be tempting because it is the additive inverse. A number and its additive inverse have a sum of 0. Choice (E) might be tempting because it is the multiplicative inverse of $\frac{1}{8}$, which is not the same number as 0.8.

4. E

This question involves the distributive property. If A is multiplied by $B + C$, the product is $(A * B) + (A * C)$.

The distributive property can be used twice in the expression you are given. Using the distribute property requires you to know how to combine terms with multiplication. After this step, you need to be able to subtract algebraic expressions. This will be addressed in chapter 2.

When you multiply 4 by $(3s + 4t)$, you multiply 4 by each term inside the parentheses, one at a time, and then add the products:

$4(3s + 4t) = 4(3s) + 4(4t) = 12s + 16t$

Likewise for the other part of the expression:

$2t(2s + 5) = 2t(2s) - 2t(5) = 4ts - 10t$

So $4(3s + 4t) - 2t(2s + 5) =$

$[4(3s) + 4(4t)] - [2t(2s) - 2t(5)] =$

$(12s + 16t) - (4ts - 10t) =$

$12s - 4ts + 6t$

So (E) is correct. Remember that terms don't share all of the same variables cannot be combined. You might get an expression like the one in (B) if you got 10s by subtracting 2ts from 12s to get 10s, but that is not allowed.

5.   A

The associative property, covered in chapter 2, allows you to "reorganize" certain expressions. It holds that $A + (B + C) = (A + B) + C$. This is like saying that if you added 5 to 7 + 3, you get the same result as you would by adding 5 + 7 to 3. The associative property also applies to expressions involving multiplication.

The key here is to recognize that the expression provided here fits the form of one the expressions in the formula given above. If $A + (B + C) =$ $8 + (4e + 9)$, then $(8 + 4e) + 9$. So (A) is the answer.

The expression you are given actually equals the one given in (B), but it does so by the commutative property, to be discussed in chapter 2. Choice (C) also contains something equal to the expression you are given, but it is the expression simplified. It has nothing to do with the associative property.

6.   B

Finding the values of algebraic expressions like this is just a matter of substituting the values of the variables you are given, and simplifying. You can substitute 7 and 4 for a and b, respectively:

$3a + 5 - 2b = 3 * 7 + 5 - 2 * 4$

Following the rules of order of operations, you can begin to simplify by carrying out multiplication. Since 3 * 7 = 21 and 2 * 4 = 8,

3 * 7 + 5 – 2 * 4 = 21 + 5 – 8

Carry out the rest of the operations, from left to right:

21 + 5 – 8 = 26 – 8 = 18

So (B) is the correct answer. Note that only the value of *a* is multiplied by 3. If you first add 5 to *a* and then multiply the sum of 12 by 3, you would probably go on to get 28, choice (C), as an incorrect answer. If you first evaluate *a* + 5 – 2*b* and multiply its value of 4 by 3 to get 12, you would incorrectly arrive at (A).

7.   C

The multiple of a given term is the product of it and a variable and/or positive integer. The least common multiple (LCM) of a pair of terms is the one that is a multiple of both, and is lower than all other common multiples. The LCM is a concept related to factoring.

You can always test each option to see whether it is a multiple of both 6 and 8*x*. 24*x* = 6 • 4*x* and 24*x* = 8*x* • 3. 48*x* is also a common multiple of 6 and 8*x*.

To find the LCM of two terms, find the prime factors of each. The prime factors of 6 are 2 and 3. The prime factors of 8*x* are 2, 2, 2, *x*, since these four terms have a product of 8*x*. The LCM of 6 and 8*x* is then the product of 2, 2, 2, *x*, and 3. The LCM multiple of two terms is the product of their **unique** prime factors. We'll go into more detail in chapter 3. The product of the unique prime factors is 24*x*, and so Choice (C) is the correct answer.

You might have arrived at (D), 48*x*, if you took the LCM to be the product of the terms. The LCM often happens to be the product, but there are many cases where the LCM is less than that. Choice (A) would be correct if you were looking for the greatest common factor of 6 and 8*x*. The greatest common factor is a related concept.

8.   C

Adding fractions requires you to find a common denominator. Since you can only add fractions with denominators of the same value, you need to find fractions equivalent to the ones being added. To do this, find the least common multiple of the denominators. Take the unique prime factors of 9 and 12: 3, 3, 2, and 2. Since their product is 36, that is the LCM of 9 and 12. Since $36 = 9 \cdot 4$, you need to multiply the top and bottom of $\frac{2x}{9}$ by 4 to find that $\frac{2x}{9} = \frac{8x}{36}$. Since $36 = 12 \cdot 3$, you need to multiply the top and bottom of $\frac{3x}{12}$ by 3 to find that $\frac{3x}{12} = \frac{9x}{36}$. $\frac{8x}{36} + \frac{9x}{36} = \frac{17x}{36}$. So (C) is the correct answer.

Choice (A) is just the result of adding the numerators and the denominators, but this not allowed. You could arrive at the incorrect (D) by multiplying the top and bottom of $\frac{2x}{9}$ by 2 instead of 4.

9. D

This question asks you to simplify a radical. The radical is the positive square root of an expression. To simplify a radical, you need to get the highest value possible on the outside of the radical sign, and the lowest value remaining inside. You can start by factoring the expression piece by piece, looking for perfect square factors.

One factor of $50x^2$ is x^2, which is a perfect square. The square root of x^2 is x, and so you can put x on the outside:

$$\sqrt{50x^2} = \sqrt{50 \cdot x^2} = x\sqrt{50}$$

Now, 2 and 25 are factors of 50. Since 25 is a perfect square, you can factor that out of the radical as well:

$x\sqrt{50} = x\sqrt{25 \cdot 2} = 5x\sqrt{2}$. Since 2 is a prime number, there is nothing left to factor.

(D) is the correct answer. Choice (E) is what you would get if you took $50x$ to be the square root of $50x^2$.

10. E

The sign $\sqrt[3]{}$ is used for the "cube root" of an expression. Cube roots are covered along with radicals in Chapter 4. You're asked to find the cube root of $-64x^{27}$, which is the expression that gives you $-64x^{27}$ when cubed. It helps to break an expression like this down into factors, and then find the root of each one. The cube root of -64 is -4; $(-4)^3 = (-4*-4)*-4 = 16 * -4 = -64$.

The cube root of x^{27} is x^9; $(x^9)^3 = (x^9 * x^9)* x^9 = (x^{9+9})* x^9 = x^{18} * x^9 = x^{18+9} = x^{27}$.

So the cube root of $-64x^{27}$ is the product of these cube roots, -4 and x^9, and which is $-4x^9$.

(E) is the correct answer. You might have chosen an answer with an exponent of 3, such as (B) or (C), if you took the cube root of x^{27} to be x^3. Even though 3 is the cube root of 27, you don't find the cube root of an exponent by taking the cube root of its power. (D) might be tempting if you take the cube root of -64 to be the number that you would double three times to get -64.

11. C

Here you are asked to solve an equation. This is one of the most important topics in Algebra I. We'll review it in detail in chapter 5. The key is to use operations in order to get the variable alone on one side of the equation. It is important to do this in the right order.

Starting with $4x + 7 = 51$, we can subtract 7 from both sides, meaning that we subtract 7 from $4x + 7$, and from 51. By performing the same operation on both sides, you guarantee that the results on each side are still equal.

$$4x + 7 = 51$$
$$\underline{-7 \quad -7}$$
$$4x = 44$$

Since the algebraic expression in the equation is a variable x being multiplied by 4, you can get the variable alone by dividing both sides of the equation by 4.

$$4x = 44$$
$$/4 \quad /4$$
$$x = 11$$

(C) is the correct answer. You might have reached (A) by dividing 51 by 4, rounding to the nearest integer, and then subtracting 7. Or you might have reached (D) by adding 7 to the right side instead of subtracting, and rounding up after dividing 58 by 4.

12. D

When asked for the solution of an expression, you need to find the value of the variable that makes the equation true. So you need to find the value of x. The goal is to get the variable alone on one side of the equation.

Start by subtracting 52 from both sides.

$$52 - 3y = 16$$
$$-52 \qquad -52$$
$$-3y = -36$$

Now, since the variable is multiplied by -3, divide both sides by that number.

$$-3y = -36$$
$$/-3 \qquad /-3$$

-36 divided by -3 equals 12, so

$$y = 12$$

So (D) is the correct answer. You might have reached (B) if you didn't realize that the result of dividing a negative number by another negative number is a positive number.

13. D

Coordinate geometry is a major algebra I topic that we'll introduce in Chapter 6. The "origin" on the coordinate grid is the point where the x axis and y axis intersect. It is right in the center of the grid. The coordinates of the origin are $(0, 0)$.

The first coordinate in the pair is the x–coordinate. If the point on the grid is to the left of the origin, the x–coordinate is positive. If it's to the right, the coordinate is negative. The second coordinate in the pair is the y–coordinate. If the point on the grid is above the origin, the y–coordinate is positive. If it's below the origin, the coordinate is negative.

Since the point is two units to the right of the origin and three units below it, the coordinates are $(2, -3)$. Choice (B) is incorrect because it has the x– and y– coordinates switched. (D) is the correct answer. (C) would result from taking points to the right of the y–axis to have negative x–coordinates instead of positive ones. And (E) would result from taking points below the *y*–axis to have positive *y*–coordinates instead of negative ones.

14. A

When you move a given number of points to the right of a point, you subtract that number from the original *x*–coordinate. When you move up from a point, you add the number of units to the *y*–coordinate. So you need to take the pair of coordinates (3, –6), and subtract 4 from 3 and add 8 to –6. You can visualize this on the grid, starting at (3, –6), moving left, and then up:

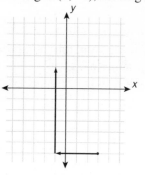

The new coordinates are (–1, 2), so (A) is correct. The pairs of coordinates in the other answer choices all result from moving in an opposite direction, or from applying each change to the wrong coordinate. (C), for instance, results from moving four units to the right of and eight units below (3, –6). (E) is moving eight units to the right and four units above (3, –6).

15. C

The slope of a line tells you its direction and ***steepness***. The equation $y = \frac{x}{4} - 5$ is in what is call "slope–intercept" form. That is an equation of the form $y = mx + b$, where *m* is the slope of the equation. Since the equation involves *x* being divided by 4, that amounts to *x* being multiplied by $\frac{1}{4}$. So the equation is $y = \frac{1}{4}x - 5$, and so $m = \frac{1}{4}$, and (C) is the correct answer.

Be careful not to confuse *m* and *b*. Choice (A) is the result of mistaking the *y*–intercept for the slope. If you figured that the value of the slope is some combination of *m* and *b*, you may have picked (B) or (D).

16. C

Recognizing graphs of linear equations is another major topic covered in chapter 7. The equation is in the standard slope–intercept form $y = mx + b$, where m is the slope and b is the y–intercept. So the equation indicates that the graph has a slope of 3 and a y–intercept of -1. A y–intercept of -1 means that the graph will cross the y–axis (the vertical axis) one unit below the x–axis (the vertical axis). The slope of 3 tells us two things. First, since the slope is positive, the graph goes up as it moves right (graphs with negative slopes go down as they move right). Second, a slope of 3 means that for every unit the graph moves right or left, it moves three units up or down, respectively.

The graph shown in (C) is the only one that has all of these features, and so it is correct. (B) would result from mistaking a negative slope for a positive one. (E) would be tempting if you confused the x–coordinates and y–coordinates. That graph has an x–intercept of -1 (instead of a y–intercept of -1), and it has a slope of $\frac{1}{3}$ instead of 3.

17. B

This question involves a system of equations. Chapter 8 covers this topic. One way to solve a system of two equations is to first solve one of the equations for one of the variables. "Solving an equation for a variable" is covered in Chapter 5.

Take the equation $8x - 2y = 14$ and solve it for y. Getting the y variable term on one side of the equation, and then dividing both sides by 2 gives us $y = 4x - 7$.

Now that we have solved one equation for y, we can substitute its value in the other equation. So we take $6x + 7y = 53$, and substitute $4x - 7$ for y:

$$6x + 7(4x - 7) = 53$$

Now we can simplify the equation and solve it for x:

$$6x + 28x - 49 = 53$$
$$34x = 102$$
$$x = 3$$

So (B) is the correct answer. You might have selected (D) if you had reported the value of y.

18. B

Another method of solving systems of equations involves combining the equations with addition or subtraction. The method works best when the variables are not multiplied by anything, and when the terms line up when you set things up for combination:

$$x - y = 17$$
$$+ \quad \underline{x + y = 63}$$
$$2x = 80$$

We're left without a y–term, because $-y$ and y have a sum of 0.

So $x = 40$. Since $x + y = 63$, $y = 23$. So (B) is the correct answer.

Choice (D) is the value of x, and (E) is the result if you didn't add the x variables and got $x = 80$ instead of $x = 40$. You would then get a value of 63 for y after solving $x - y = 17$.

19. D

This is a linear inequality, dealt with in chapter 9. Solving inequalities like this is as straightforward as solving equations. In fact, it usually helps to begin by pretending that you're solving an equation. It can get more complicated than that, as you'll see. In this case, start by adding 6 to both sides of the inequality.

$$2x - 6 > 8$$
$$\underline{+6 \ +6}$$
$$2x > 14$$

Dividing both sides by 2 gives you the solution, $x > 7$, so (D) is the answer. You would have gotten (C) by mistakenly subtracting 6 from the right side instead of adding, or (E) by mistakenly dividing 8 by 2 first and then adding 6 to both sides.

20. A

Solving this inequality is a bit complicated. First, it involves two variables. Since all the answer choices are solutions in terms of y, your goal is to get y alone on one side of the inequality. Subtracting $4x$ from both sides gives us $-9y \geq -12 - 4x$.

To get **y** alone on the left side, we need to divide both sides by –9. But when you divide by a negative number, the direction of the inequality changes! This is very important. A lot of wrong test answers come as a result of forgetting this step! You might have selected (E) if you overlooked it.

So when you divide both sides by –9, the "greater than" sign becomes a "less than" sign, and the solution is $y \le \frac{4x+12}{9}$. Choice (A) is the answer.

21. E

Multiplication of polynomials is covered in chapter 10. Using the FOIL method (multiplying the First, Outside, Inside, Last terms), you can find the product by multiplying each term inside the first set of parentheses with each term inside the second set.

$$(4x^2 + 2x)(3x + 2) =$$

Then add those four products.

$$4x^2 * 3x = 12x^3$$
$$4x^2 * 2 = 8x^2$$
$$2x * 3x = 6x^2$$
$$2x * 2 = 4x$$
$$12x^3 + 8x^2 + 6x^2 + 4x = 12x^3 + 14x^2 + 4x$$

(E) is correct. (C) and (D) result from leaving out one of the x^2 terms when adding. Be sure to perform each multiplication, and add all of the results.

22. B

As chapter 10 explains, you can add polynomials by combining the terms that have the same powers. So you can add $3x^2$ and $4x^2$, but you can't add $4x^4$ and $5x^3$. Use the associative property to arrange the terms in an order that makes adding easier:

$$(4x^4 + 5x^3) + (3x^2 + 4x^2) + (2x + 3)$$

It turns out that $3x^2$ and $4x^2$ are the only two terms that can be combined. Their sum is $7x^2$.

So the sum of the polynomials is $4x^4 + 5x^3 + 7x^2 + 2x + 3$, and (B) is the correct answer.

Be careful when matching up like terms. (D) is the result of trying to add $4x^4$ to $5x^3$ and $2x$ to 3. You might have gone with (A) if you avoided that first mistake, but not the second one.

23. B

Factoring is another important topic related to polynomials. Polynomials such as $x^2 - 7x + 10$ can often be factored into a pair of binomials like the ones found in the answer choices. Though you need only to select one factor to answer the question, you need to find both factors to be sure you have found one that works.

Factoring polynomials can involve some trial and error. If $x^2 - 7x + 10$ has factors $(x + j)$ and $(x + k)$, there is a straightforward way of finding j and k.

Since $yk = 10$, j and k must be both positive or both negative. The middle term is $-7x$, so j and k must both be negative. $(x + j) * (x + k) = x^2 - (j + k) x + jk$, so j and k must have a sum of -7 and a product of 10. Start by listing the factors of 10, both negative and positive. See if you can find a pair that has a sum of -7:

Factors of 10: $\{-10. \underline{-5}. \underline{-2}, -1, 1, 2, 5, 10\}$

We underlined -5 and -2 because they add up to -7 and have a product of 10. So j can be -5 and k can be -2. That means the factors of $x^2 - 7x + 10$ are $x - 5$ and $x - 2$. One of those is choice (B), which is the correct answer. (C), (D), (E) each give a value of j that is a factor of 10, but there is no value of k where j and k would have a product of 10 and a sum of -7.

24. C

This problem requires you to solve a quadratic equation. You can use factoring to get to the solution, but you also need to know how to use the factors to get the final answer.

If $x^2 + 6x - 16$ has factors $(x + j)$ and $(x + k)$, then $(x + j) * (x + k) = x^2 - (j + k)x + jk$. So j and k must have a sum of 6 and a product of -16.

Start by listing the factors of -16, both negative and positive. See if you can find a pair that has a sum of 6 and a product of -16:

Factors of -16: $\{-16, -8, -4, \underline{-2}, -1, 1, 2, 4, \underline{8}, 16\}$

-2 and 8 are underlined because they meet the conditions.

The factors of $x^2 + 6x - 16$, then, are $x - 2$ and $x + 8$. So if $x^2 + 6x - 16 = 0$, then $(x - 2)(x + 8) = 0$. Since the product of the factors is 0 when $x - 2 = 0$ or $x + 8 = 0$, $x = 2$ or $x = -8$.

So the solutions are not j and k, but their additive inverses. If you overlooked that step, you might have picked (A). Choice (C) is the correct answer. The other choices result from incorrect factoring.

25. C

Here you need to solve a quadratic equation, which can often be solved through factoring. Seeing that the answer choices involve radicals, you should realize that it will be difficult to find the factors through trial and error. Instead, use the quadratic formula to find the solutions. For the equation $ax^2 + bx + c = 0$:

$$x = \frac{-b \pm \sqrt{b^2 - 4ac}}{2a}$$

If $2x^2 + 6x + 3$, then $a = 2$, $b = 6$, and $c = 3$. Plug those values into the formula:

$$x = \frac{-6 \pm \sqrt{6^2 - 4(2)(3)}}{2(2)}$$

$$= \frac{-6 \pm \sqrt{36 - 24}}{4}$$

$$= \frac{-6 \pm \sqrt{12}}{4}$$

$$= \frac{-6 \pm \sqrt{(4)(3)}}{4}$$

$$= \frac{-6 \pm 2\sqrt{3}}{4}$$

$$= \frac{-3 \pm \sqrt{3}}{2}$$

Choice (C) is the answer. You would have gotten (D) if you forgot to divide $2\sqrt{3}$ by 2 in the final step.

26. A

A fraction is ***undefined*** if its denominator has a value of 0, as division by 0 is defined. So the fraction $\frac{3x-6}{4x+12}$ is undefined when $4x + 12 = 0$ as division by 0 is undefined. Finding the answer, then, is just a matter of solving that equation:

$$4x + 12 = 0$$

$$\underline{-12 \quad -12} \text{ [Subtract 12 from both sides]}$$

$$4x = -12$$

$$\underline{/4 \quad /4} \text{ [Divide both sides by 4]}\_$$

$$x = -3$$

Choice (A) is correct. You might have picked (D) if you mistakenly added 12 to the right side of the equation instead of subtracting. (C) results from working with the numerator instead of the denominator and solving $3x - 6 = 0$.

27. C

The question asks you to simplify a fraction with variables in the numerator and denominator. Simplifying it is a matter of dividing the numerator by the denominator. Since the numerator and denominator are each the product of multiplication, you could combine factors from the top and bottom to simplify the fraction.

Starting with the numbers, note that 24 divided by 6 is 4, or $\frac{24}{6} = \frac{4}{1}$. So:

$$\frac{24x^6 y^3 z^8}{6x^2 yz^4} = \frac{4x^6 y^3 z^8}{x^2 yz^4}$$

Take the variable *x* terms next:

$$\frac{x^6}{x^2} = x^{6-2} = x^4 \text{, so } \frac{24x^6 y^3 z^8}{6x^2 yz^4} = \frac{4x^4 y^3 z^8}{yz^4}$$

Take the variable *y* terms:

$$\frac{y^3}{y} = y^{3-1} = y^2 \text{, so } \frac{4x^4 y^3 z^8}{yz^4} = \frac{4x^4 y^2 z^8}{z^4}$$

Finally, take the *z* terms:

$$\frac{z^8}{z^4} = z^{8-4} = z^4 \text{, so } \frac{4x^4 y^3 z^8}{yz^4} = 4x^4 y^2 z^4$$

The fraction is now simplified, and (C) is the answer. It is important to remember to subtract the powers when dividing exponents. (B) would result from incorrectly dividing those powers.

28. A

This is an algebraic word problem. This one doesn't involve distances, as many standard rate problems do, but you can treat the volume of water like a distance. A standard formula is:

Rate * Time = Distance, or $rt = d$

We're given the rate at which the faucet would fill the pot (3 fluid ounces per second), and the ***distance***, which is 48 fluid ounces. The unknown amount is the time it takes to fill the pot at this rate. So you can fill in the values you have to get an equation which you can solve.

If $r = 3$ and $d = 48$, then $3t = 48$. To solve this equation for t, all you have to do is divide both sides of the equation by 3. That gives you $t = 16$, making (A) the correct answer. (D) would be the result of subtracting 3 from the right side of the equation instead of dividing it by 3.

29. C

This word problem involves percentages, which appear frequently in real world algebraic problems. You aren't being asked here to find the percentage of a number; you're being asked to figure out that number.

Call the price of the price of the meal x. The total amount of money Darren gave, minus the $0.52 sales tax should equal the price of the meal x, plus a 15% tip, which is 15% of x or $0.15x$. So

$x + 0.15x = 2.48 - 0.52$

We can simplify each side of the equation to get

$1.15x = 11.96$

To find the value of x, divide both sides by 1.15

$x = \dfrac{11.96}{1.15} = 10.40$

So the price of the meal was $10.40. (C) is the correct answer. Choice (A) might be reached by taking 15% off of 11.96, or (D) by adding the sales tax to the final cost instead of subtracting, and then dividing by 1.15.

30. E

Algebra questions involving areas and perimeters are very common. The area of a rectangle is the product of its length and width. The perimeter is the sum of the measures of all for sides. So if x stands for length and y stands for width, use the following formulas:

Area $= x * y$

Perimeter $= 2x + 2y$

In this problem, then, you actually have to solve a system of equations (Chapter 8):

$$x * y = 168$$
$$2x + 2y = 52$$

You could start by solving the perimeter equation for y.

$$2y = 52 - 2x$$
$$y = 26 - x$$

Now that you've solved one equation for y, substitute that solution for y in the other equation:

$$x * y = x(26 - x) = 168$$
$$26x - x^2 = 168$$

It looks like you'll need to solve a quadratic equation (Chapter 11). Get it into the standard form and solve:

$$x^2 - 26x + 168 = 0$$
$$(x - 12)(x - 14) = 0$$
$$x = 12 \text{ or } x = 14$$

So the length is 12 or 14. If you solve one of the equations for y after plugging in a solution, you'll get the other number for the width. The dimensions are 12 ft. and 14 ft. then, and (E) is correct. The dimensions in (A) and (B) work out to the right area but not to the right perimeter, and the dimensions in (C) and (D) work out to the right perimeter but not to the right area.

Question Number	Topic	Chapter in Which the Topic is Covered
1	Order of Operations	1
2	Fractions and Decimals	1
3	Inverses	1
4	Simplifying Algebraic Expressions	2
5	Associative Property	2
6	Evaluating Algebraic Expressions	2
7	Factors and Least Common Multiplies	3
8	Common denominators	3
9	Radicals	4
10	Cube Roots	4
11	Solving Equations	5
12	Solving Equations	5
13	Identifying Coordinates	6
14	Plotting Points	6
15	Slope and Linear Equations	7
16	Graphing Slope-Intercept Form Equations	7
17	Solving Systems of Equations	8
18	Solving Systems of Equations	8
19	Solving Linear Equalities	9
20	Solving Linear Equalities with Division by Negative Numbers	9
21	Multiplying Binomials	10
22	Adding Polynomials	10
23	Factoring Trinomials	10
24	Solving Quadratic Equations	11
25	Solving Quadratic Equations	11
26	Undefined Algebraic Fractions	12
27	Simplifying Algebraic Fractions	12
28	Rate Problems	13
29	Percentage Problems	13
30	Area Problems	13

Number Properties

WHAT ARE NUMBER PROPERTIES?

There are certain basic features of numbers you'll need to be familiar with in order to get very far in Algebra I. These properties don't apply to just individual numbers; they also apply to *expressions*. Much of what we'll cover in this chapter deals with basic expressions of arithmetic. We won't introduce topics unique to algebra until we get to chapter 2.

The explanations in the following sections will help you to follow the walkthroughs of the five key questions we'll tackle later in the chapter. The material in those sections, however, will prove to be important for virtually every topic in the rest of the book.

CONCEPTS TO HELP YOU

Let's begin by reviewing some of key concepts behind number properties. All of these are essential for handling the common question types later in the chapter. You should come back this section whenever you are unclear about one of those questions or one of the chapter quiz questions. Since algebra involves expressions including positive and negative integers, decimals, fractions, and absolute values, we need to go through the properties of those things. It will also be helpful to know about some of the special properties of 0, 1, −1, and inverses.

Expressions

You are going to hear the words expression, operation, and evaluate a lot in this book. An *expression* is a combination of numbers, variables, and at least one operation. In chapter 1, we'll stick to numbers and operations, and we'll get to variables in chapter 2. An *operation* is something you perform on one or more numbers to obtain a value. Among the more common operations are addition, subtraction, multiplication, and division. To evaluate an expression is to find its value. The value of the expression $1 + 2$ is 3. For now, that's all there is to it.

Evaluating expressions of arithmetic is a major focus of this chapter. This is a skill you need to master in order to tackle algebra I.

Integers, Fractions, and Decimals

Let's define these common numbers types and then see how we can work with them:

> **Integers** consist of the whole numbers and their opposites $\{...-3, -2, -1, 0, 1, 2, 3...\}$.

> **Fractions** are numbers that can be expressed with two integers, a numerator and a denominator. $\frac{1}{2}$ is a fraction with a numerator of 1 and a denominator of 2. You might think of this fraction of representing one part out of 2. Suppose that a pizza pie is cut into 8 slices of equal size, and 3 of them are taken. You could say that $\frac{3}{8}$ of the pizza, or 3 parts of 8, have been taken.

> **Decimals** are a special kind of fraction, really. Money values measured in cents are one of the most common examples of decimals. 25 cents can be written out as $0.25. That amount equals $\frac{2}{10}$ of a dollar plus $\frac{5}{100}$ of a dollar. Each place in a decimal, as you go from left to right, represents a smaller fraction amount. The decimal 0.5639 equals $\frac{5}{10}$ plus $\frac{6}{100}$ plus $\frac{3}{1,000}$ plus $\frac{9}{10,000}$. Notice that each denominator is ten times greater than the previous one.

We will explain later on how to convert numbers from fractions to decimals, and vice versa.

Positive and Negative Numbers

Imagine a number line such as the one below:

Number lines are often drawn with the number 0 in the center, but it is not always necessary. As you go right on the number line from 0, you see increasing positive numbers. As you go left, you see decreasing negative numbers. Don't be fooled by the fact that the negative numbers may seem to increase as you move left. The *farther* a number is to the left of 0 on the number line, the *lower* its value. So –5 is less than –4 and –3 is less than –2.

DOUBLE NEGATIVES

You'll often see numbers with two negative signs. Such numbers are in fact positive. We will often use parentheses when writing such "double negatives":

$-(-5) = 5$

$-(-10) = 10$

You could say that the two negative signs "cancel each other out."

Properties of the Numbers –1, 0, and 1

Adding by 0

The sum of 0 and any number is that number.

$2 + 0 = 2, 3 + 0 = 3$

Multiplying by 0, 1, and –1

For *any* number, the following rules hold. Let's call our number A:

$A \times 0 = 0$

$A \times 1 = A$

$A \times -1 = -A$

Now, what if A is negative? Is $-A$ a positive number or a negative one? The answer is *positive*. The negative of a negative number is a positive number.

Absolute Value

The number line can also help us to make sense of absolute value. Using the letter A to stand for a number, we express its absolute value with the expression $|A|$. Straight vertical brackets are used only in connection with absolute value.

Now, the value of an absolute value expression is positive, no matter whether the expression inside the brackets is positive or negative. In fact, the absolute value of any number is its distance from 0 on the number line. Let's take two numbers, 3 and –3:

Each of those numbers is three units away from 0. They each have the same, positive, distance from 0. After all, have you ever heard of a negative distance?

So $|-3| = 3$ and $|3| = 3$.

ANOTHER WAY TO LOOK AT IT

If A is positive, then $|A| = A$;

If A is negative, then $|A| = -A$ which is positive.

In either case, $|A|$ is positive.

Inverses

You won't need to use inverses to evaluate expressions, but multiplicative and additive inverses are number properties that will be important to know for topics later in this book.

Multiplicative Inverses

A number and its multiplicative inverse have a product of 1. To get the multiplicative inverse of a fraction, you only need to "turn the fraction upside down", that is, swap the numerator and the denominator. So the multiplicative inverse of $\frac{2}{3}$ is $\frac{3}{2}$ The product of these fractions is $\frac{6}{6}$, and any fraction with a numerator and a denominator that are equal has a value of 1.

Also note that this approach works with integers as well. Every whole number can be rewritten as a fraction with a denominator of 1. When you see that $8 = \frac{8}{1}$, it makes it easier to recognize $\frac{1}{8}$ as the multiplicative inverse of 8.

Multiplicative inverses are also called reciprocals.

Additive Inverses

The additive inverse of a number gives you a sum of 0 when combined with that number. To get it, just add a negative sign to the number in question. If your original number was positive, the additive inverse is the number with a negative sign in front. Suppose your original number is negative: let's take −5. The additive inverse of that number is −(−5); the negative signs cancel out, and you are left with 5. You can also look at the additive inverse as the product of a number and −1. Additive inverses are also called opposites.

STEPS YOU NEED TO REMEMBER

The concepts we just reviewed are essential to arithmetic at the Algebra I level. When we use them, we also want to follow some basic steps. As you will see, there are rules that lay down the order in which you go about evaluating expressions. There are also important steps to follow when you are dealing with numbers in various forms, such as fractions and decimals. Finally, there are certain steps you can take to be sure that you are carrying out operations correctly.

Order of Operations

In order to evaluate an expression, we need to carry out all of its operations. How do you go about that? You are not supposed to just go from left to right in an expression. There is a certain *order of operations* to follow.

To evaluate an expression, perform operations in the following order.

1. Perform operations inside parentheses

2. Do all powers (we'll get to powers in Chapter 4)

3. Do all multiplication and division operations, one at a time, going from left to right.

4. Do all addition and subtraction operations, one at a time, going from left to right.

You might be asking "why this order?" and "why not the reverse order?" We can't get into that here. These are just rules, like the rules of grammar or driving on the right side of the road. They could have been different, but it's important that we follow them as they have been set.

Converting Numbers

Converting Fractions to Decimals

Converting a fraction to a decimal is straightforward: Just divide the *numerator* (the top number) by the *denominator* (the bottom number). So the decimal value of $\frac{1}{4}$ is 1 divided by 4. If you can't use a calculator, here's how you would do it on paper:

$$\begin{array}{r} 0. \\ 4\overline{)1.00} \end{array}$$ Set it up the division like this, including the decimal place after the 1, followed by several zeroes; treat it as though you are dividing 100 by 4. Just remember to put a decimal place in front of the result.

$$\begin{array}{r} 0.2 \\ 4\overline{)1.0^2 0} \end{array}$$ Now divide 10 by 4. This is what you would do as a first step in dividing 100 by 4. Your result is 2, with a remainder of 2. The little 2 you see between the two zeroes is that remainder. You need that remainder for the next division step.

$$\begin{array}{r} 0.25 \\ 4\overline{)1.0^20} \end{array}$$ That next step is dividing 20 by 4. How did we get 20? You take the number remainder and the next digit to get a 2–digit number, which you then divide by 4. 20 divided by 4 is 5, with a remainder of 0. Once you do a division that gets a remainder of 0, you're finished.

Many conversions will take more work, but the principle is always the same.

Converting Decimals to Fractions

If the decimal you are given to work with is *non–terminating* (meaning that it doesn't go on forever), you can always convert it to fraction with a whole number *numerator* and a denominator that is a power of 10.

Take 0.123. The first digit to the right of the decimal point, 1, is the *tenths* place; the next digit is in the *hundredths* place; and the last digit is in the *thousandths* place. Take the last place of the decimal. The numbers on the right of the decimal point stand for that many thousandths. So 0.123 is 123 thousandths, or $\frac{123}{1,000}$.

Here are other examples.

0.0024 is 24 ten–thousandths, or $\frac{24}{10,000}$

0.67 is 67 hundredths, or $\frac{67}{100}$

0.000035 is 35 millionths, or $\frac{35}{1,000,000}$

Many of the fractions you'll get by converting decimals can be *simplified*. A simplified fraction is one that is rewritten as a fraction with the same value, but with a smaller numerator and a smaller denominator. We'll get to this later, in chapter 3.

Combining Numbers

Subtracting Numbers

Take the number 2. When you add 3, you get 5. Think of adding 3 as moving that many notches to the right on the number line. By moving three notches to the right of 2, you wind up at 5.

Suppose you want to subtract 4 from 5. Move 4 notches left from 5, and you wind up at 1. That makes enough sense, since $5 - 4 = 1$.

You might not need a number line to deal with basic addition and subtraction, but it is useful when explaining how to subtract a larger number from a smaller one.

What is $3 - 7$? Start at 3, and move 7 notches left.

This shows that $3 - 7 = -4$. So it helps to explain another way of answering this kind of question: to subtract 7 from 3, find $7 - 3$, and add a negative sign to the difference.

Subtracting from Negative Numbers

Adding to or subtracting from negative numbers can be more confusing, but you can use the number line once again to manage those operations. What is $-2 + 4$? -6 might be a tempting answer on a multiple–choice test, but it would be incorrect. Instead of adding 4 to 2 and sticking on a negative sign, use the number line. Since you are adding, start at -2 and move four notches right. So $-2 + 4 = 2$.

Let's look at a question where a negative number is subtracted from another negative number.

What is $-1 - (-3)$? Using the number line, we can find that the answer is 2.

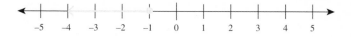

Multiplying and Dividing Negative Numbers

It would be difficult to use the number line to explain multiplication and division of negative numbers. When multiplying or dividing of two numbers, find the product or quotient as though no negative numbers were involved. Now, if both numbers involved are negative, the result is a positive number. If one of the two numbers is negative, the result is negative.

$$-4 \times -8 = 32$$
$$-3 \times 5 = -15$$
$$6 \times -2 = -12$$
$$16 \div -8 = -2$$
$$-28 \div -7 = 4$$

Why does multiplying two negative numbers, for instance, result in a positive product? Remember that a *double negative* equals a positive, such that $-(-4) = 4$. If you multiply 6 by -7, you add a negative sign to the product of 6 and 7. In multiplying with a second negative, you add a second negative sign. The two signs then cancel each other out. So you could think of the product of -6 and -7 as $-(-42)$, which equals 42.

Multiplying Fractions

To multiply two fractions, simply multiply the numerators and the denominators.

$$\frac{1}{2} \times \frac{3}{4} = \frac{1 \times 3}{2 \times 4} = \frac{3}{8}$$

Here's a helpful way to think about multiplying fractions: you're essentially looking for a fraction *of* a fraction. When you evaluate $\frac{1}{2} \times \frac{3}{4}$, you find the value of one half of $\frac{3}{4}$. It turns out that one half of $\frac{3}{4}$ is $\frac{3}{8}$.

Dividing Fractions

In order to divide fractions, you need to know how to multiply them. That's because you can divide by a fraction by multiplying by the "multiplicative inverse" of a fraction.

> Dividing a number by a fraction is the same as multiplying by the multiplicative inverse of the fraction. The result of dividing any number by itself is 1. So take the fraction $\frac{2}{5}$. Recall that to divide a number by $\frac{2}{5}$, you multiply it by the inverse, $\frac{5}{2}$. So
>
> $$\frac{2}{5} \div \frac{2}{5} = \frac{2}{5} \times \frac{5}{2} = 1$$

Adding and Subtracting Fractions

To combine two fractions with addition or subtraction, they must have the same value in the denominator. To combine such fractions, just add or subtract the numerators. The value of the denominator stays the same.

$$\frac{4}{9} + \frac{3}{9} = \frac{4+3}{9} = \frac{7}{9}$$

$$\frac{8}{11} - \frac{5}{11} = \frac{8-5}{11} = \frac{3}{11}$$

If the two fractions don't have the same denominator, don't give up. You can rewrite one or both of them so that they have the same denominator. "Finding a common denominator" is a very important concept, and we'll deal with it in chapter 3.

Multiplying and Dividing Decimals

You can multiply decimals just as you would whole numbers, as long as you take a couple of important extra steps:

1. Count the number of decimal places in each number.

2. Add these numbers together.

3. Take the product of the numbers and add a decimal point, so that the number of decimal places of the product is the number you got in step 2.

So what is the product of 0.25 and 0.35? Start by multiplying the whole numbers 25 and 35: $25 \times 35 = 875$. Since there are two decimal places in both 0.25 and 0.35, you need to add a decimal point so that the product has four decimal places. Count the number of decimal places from the right to the left and insert the zero as show to get four decimal places. If you do this, you get a product of 0.0875.

To divide decimals, take a similar approach, with one key difference. After dividing the numbers (ignoring the decimal points, and treating them like whole numbers), subtract the number of decimal places in the divisor (the number you are dividing by) from the number of decimal places in the dividend (the number you are dividing). The difference is the number of decimal places in your result.

What is 0.35 divided by 0.7? Divide 35 by 7 to get 5. The dividend, 0.35 had two decimal places, and the divisor, 0.7, has one decimal place. Since $2 - 1 = 1$, you should insert a decimal into your result so that there is one decimal place. So 0.35 divided by 0.7 is 0.5.

Adding and Subtracting Decimals

Adding and subtracting decimals is a lot like adding and subtracting whole numbers; you just need to keep track of decimal places.

When adding two decimals, say, 0.84 and 0.8, there are three things you will want to do:

1. Arrange the numbers so that the decimal points line up.

2. Add zeroes to the end of one of the numbers so that each decimal has the same number of places.

3. Make sure that you have a decimal point for the sum of the numbers, lined up with the other decimal points.

So to set up the addition of 0.84 and 0.2 before actually adding anything:

```
 0.84
 0.20   [we added a zero to the end of 0.2 (the value is still the same)]
   .     [we lined up all of the decimal points, including the one
          we put in for the sum]
```

Now we can add the decimals just like they are whole numbers, working around the decimal point we put in place:

0.84
<u>0.20</u>
1.04

It's like adding 84 and 20 to get 104 (don't forget to carry the 1 in the tens column!). We got a sum of 1.04 because of the position of the decimal place.

We have just gone through the key steps for carrying out the basic operations of arithmetic involving integers, fractions, and decimals.

STEP–BY–STEP ILLUSTRATION OF THE 5 MOST COMMON QUESTION TYPES

Now let's cut to the chase and examine common types of questions. Most of these involve evaluating expressions, but we want to focus on different number properties in each one. So we'll deal with positive integers, negative integers, zero, decimals, fractions, and absolute values. Finally, we'll examine a question related to inverses.

Question 1: Evaluating Arithmetic Expressions with Integers

Evaluate $2 + 6 \times 3 - 25$.

(A) −49

(B) −45

(C) −5

(D) −1

(E)   1

As you carry out the operations one at a time, it helps to substitute results into the expression. Now, the first two parts of the order of operations don't seem to apply here; there are no parentheses or powers to deal with. This means we can go on to the third step and carry the multiplication. Since a multiplication sign lies between the 6 and the 3, we have to multiply those numbers first.

$$6 \times 3 = 18$$
$$2 + \underline{6 \times 3} - 25 =$$
$$2 + \underline{18} - 25$$

The underlined parts of the expressions above indicate where we are carrying out an operation, and then replacing it with the result.

Now all that is left to do is add and subtract. Start at the left side of the expression, and perform one operation at a time.

$$2 + 18 = 20$$
$$\text{So } \underline{2 + 18} - 25 =$$
$$\underline{20} - 25$$

Finally, $20 - 25 = -5$, so **(C) is the answer**. As you can see, the last step involves subtraction with two positive numbers that results in a negative number. The result is negative because the second number is larger than the first.

The other choices are actually answers that would result from a number of very common mistakes. (D) is the result of ignoring the order of operations, and simply carrying out the operations from left to right. You would first get 8 as the sum of 2 and 6; then multiply 8 by 3 to get 24; and finally, subtract 25 from 24 to give you −1. Every operation here was performed correctly, but the wrong answer is reached because they were performed in the wrong order. (E) is the result of the same incorrect order of operations, as well as a subtraction mistake in the last step. Someone who just took 24 − 25 as having the same value as 25 − 24 would choose 1 as the value of the expression. (A) also involves a subtraction error in the last step. Someone who got to −49 might have tried to evaluate 25 − 24 by adding 24 and 25, and then attaching a negative sign to the sum. That same error is what might lead someone to (B). −45 is actually what you get by subtracting 25 from −20. Here, the options were performed in the correct order, but the subtraction mistake led to an incorrect value.

Question 2: Evaluating Expressions with Negative Integers and Zero

$-6 + 5 \times (3 - 3) - 9 \times 2 =$

(A) -30

(B) -24

(C) -20

(D) -19

(E) -18

Evaluating this expression involves everything covered in the first question, but it also involves some new elements. This one features parentheses, a negative number, and, as it turns out, multiplication by 0.

Remember to substitute results into the expression as you carry out the operations one at a time.

$-6 + 5 \times (3 - 3) - 9 \times 2$ has one operation within parentheses: $3 - 3$. According to the rules of the order of operations, you must carry out that operation first. Since $3 - 3 = 0$, you can replace $(3 - 3)$ with 0.

$-6 + 5 \times \underline{(3 - 3)} - 9 \times 2 =$

$-6 + 5 \times \underline{0} - 9 \times 2$

This expression has no powers, so we don't have to worry about that kind of operation. Next, we have to carry out any multiplication and division we find. There are two pairs of numbers combined by multiplication: 5 is multiplied by 0, and 9 is multiplied by 2. Find those products one at a time, going from left to right:

$5 \times 0 = 0$, so

$-6 + \underline{5 \times 0} - 9 \times 2 =$

$-6 + \underline{0} - 9 \times 2$

$9 \times 2 = 18$, so

$-6 + 0 - \underline{9 \times 2} =$

$-6 + 0 - \underline{18}$

Now we carry out addition and subtraction operations, one at a time, going from left to right:

$$-6 + 0 - 18 =$$
$$-6 - 18$$

Finally, $-6 - 18 = -24$, so **(B) is the answer.** As you can see, this question requires multiplication and subtraction involving negative numbers.

If you ignored the rule that requires you to carry out multiplication before addition, you might have gotten to the expression $-6 + 5 \times 0 - 9 \times 2$, and then added -6 and 5 instead of multiplying 5 and 0. The result would be $-1 \times 0 - 9 \times 2$; if you did everything correct from that point on, you would get -18, which is the incorrect choice (E).

If you forgot that the product of a number and 0 is always 0, and took the product of 5 and 0 to be 5, you might have gone ahead and evaluated $-6 + 5 - 9 \times 2$. If you did everything else right from that point on, you would have gotten -19, which is the incorrect choice (D). If you didn't follow the order of operations from that point, and performed the multiplication last, you would have gotten -20, which is the incorrect choice (C). Choice (A) could be the result of ignoring the order of operations once you get to the step where you have $-6 - 9 \times 2$. If you first subtract 9 from -6, you get -15. Multiplying that number by 2 gets you -30.

Question 3: Evaluating Expressions with Fractions and Decimals

What is the value of $\frac{1}{8} + 1.5 \times \frac{1}{4}$?

Note first that you're being asked the same thing that you were asked in questions 1 and 2: to evaluate an expression. Don't let the difference in the wording of the question throw you off.

You're not given answer choices here, so you need to calculate the result. You might encounter an open–ended question like this on the Grid–in section of the SAT. Such grid–in questions usually allow you give an answer in either the form of a decimal or a fraction. We'll work through the question in two ways, so you can have both a decimal and a fraction solution.

The key again is the order of operations, but this question also involves an extra step. Here, you are working with fractions and a decimal in the same expression. In order to combine these numbers, they would have to be either all fractions or all decimals. Fortunately, you can convert from one form to the other.

You might find it easier to convert the decimal to a fraction; that way, it only takes one step to get all three numbers in the same form. So let's convert 1.5 and substitute it with its fractional form.

Since $1.5 = \frac{3}{2}$, the expression can be rewritten as $\frac{1}{8} + \frac{3}{2} \times \frac{1}{4}$.

Following the order of operations rules, we must begin with the multiplication:

$$\frac{3}{2} \times \frac{1}{4} = \frac{3}{8}$$

Substituting this product into our expression, we now have to evaluate $\frac{1}{8} + \frac{3}{8}$. The sum of these fractions is $\frac{4}{8}$. When we get to factors later in the book, we'll be in a better position to explain why $\frac{4}{8}$ is the same value as $\frac{1}{2}$. Both fractions also happen to equal $\frac{2}{4}$. Any of those fractions should be acceptable answers on a test, unless you're asked for a fraction in its simplest form.

Now let's work this out the other way, by converting the fractions to decimals:

$$\frac{1}{8} = 0.125$$

$$\frac{1}{4} = 0.25$$

These relations are very important. It would help to memorize the decimal values of common fractions.

Some decimal values of common fractions are:

$\frac{1}{2} = 0.5$

$\frac{1}{4} = 0.25$

$\frac{1}{5} = 0.2$

$\frac{1}{8} = 0.125$

$\frac{1}{10} = 0.1$

$\frac{1}{3}$ has a value of approximately 0.33; in fact this is a repeating decimal, a continuous series of 3's!

Let's rewrite our expression, replacing the fractions with their decimal values: $0.125 + 1.5 \times 0.25$. Carrying out the multiplication first, $1.5 \times 0.25 = 0.375$, we have $0.125 + 0.375$. The sum is 0.5.

You could also have converted $\frac{4}{8}$ to decimal form to get to this last step.

Question 4: Evaluating Absolute Value Expressions

Evaluate $1 - \left(\left| -5 \right| \times \left| 3 - 7 \right| \right)$.

To evaluate an expression like this, you'll need to remember the other rules covered earlier, and you will have to find the absolute values contained in the expression. Start with the absolute values. Get the values for the two that are in the expression ($\left| -5 \right|$ and $\left| 3 - 7 \right|$) and substitute them in.

The absolute value of a number is its distance from 0 on the number line. To get the absolute value of a negative number, you need only remove the negative sign. So $\left| -5 \right| = 5$. As for $\left| 3 - 7 \right|$, you'll need to find the value of $3 - 7$ first. Since $3 - 7 = -4$, $\left| 3 - 7 \right| = \left| -4 \right| = 4$.

Now that you have values for the absolute value expressions, replace $|-5|$ with 5 and $|3-7|$ with -4:

$$1-\left(|-5|\times|3-7|\right)=$$
$$1-(5\times4)=$$
$$1-20=$$
$$-19$$

As you can see, evaluating the absolute values first can go a long way toward clearing the clutter from these expressions.

Question 5: Finding the Inverse

What is the multiplicative inverse of $-\frac{7}{4}$?

(A) $-\frac{4}{7}$

(B) $\frac{3}{7}$

(C) $\frac{4}{7}$

(D) $\frac{7}{4}$

(E) $\frac{11}{4}$

The product of a number and its multiplicative inverse is 1. So we need to find the number that gets us a product of 1 when multiplied by $-\frac{7}{4}$. To get the multiplicative inverse of a fraction, you need only "turn the fraction upside down"; that is, swap the numerator and the denominator. That would make $-\frac{4}{7}$ the multiplicative inverse of $-\frac{7}{4}$.

It can be tricky to see this in the case of negative numbers, but you can think of $-\frac{7}{4}$ as $\frac{-7}{4}$. Swapping the numerator and denominator, we get $\frac{4}{-7}$, which is the same as $-\frac{4}{7}$. That is why **(A) is correct.**

We can check this by multiplying $-\frac{7}{4}$ by $-\frac{4}{7}$. Handle this by using $\frac{-7}{4}$ and $\frac{4}{-7}$. That amounts to multiplying the numerators and the denominators:

$$\frac{-7}{4} \times \frac{4}{-7} = \frac{-7 \times 4}{4 \times -7} = \frac{-28}{-28}$$

Now, every fraction that has a numerator and denominator of equal value equals 1. So the product we just found equal 1. This proves that $-\frac{4}{7}$ is the multiplicative inverse of $-\frac{7}{4}$.

Let's see why the other choices are incorrect.

> When looking at (C), it might seem reasonable to think that the inverse of a negative number is a positive one. Note that the multiplicative inverse of a negative number must be another negative number. The product of a negative number and a positive number, after all, is negative. We're looking for a product of 1, a positive number.

This would be a good reason to rule out (D) and (E) as well, but they might be tempting if you have the additive inverse of $-\frac{7}{4}$ in mind. (D) presents the real additive inverse of $-\frac{7}{4}$. This is because $-\frac{7}{4}$ and $\frac{7}{4}$ have a sum of 0. Choice (E) involves some confusion about the additive inverse. If you chose it, you might have been taking the sum of a number and its additive inverse to be 1 instead of 0. Choice (B) involves a similar confusion. If you were looking for the number that gets you -1 when *added* to $-\frac{7}{4}$, you might have gone with $\frac{3}{7}$.

CHAPTER QUIZ

1. $(3 \div (5 \times (4 + 2))) =$

2. $8 \times -1 + 3 \times 0 =$
   (A) −11
   (B) −8
   (C) −5
   (D) 0
   (E) 16

3. What is the value of $8 - 3 \times (-7 - 2)$?
   (A) −45
   (B) 35
   (C) −7
   (D) 23
   (E) 27

4. $(1 - 0.06) \times 7 =$
   (A) 0.028
   (B) 0.658
   (C) 0.28
   (D) 2.8
   (E) 6.58

5. What is $-4 \times (15 \div -3 - 2)$?
   (A) −28
   (B) −22
   (C) 12
   (D) 18
   (E) 28

6. Which expression has a value of 72?
   (A) $-4 \times -9 \times -2$
   (B) $9 - 6 \times 24$
   (C) $-9 - 9 \times -9$
   (D) $12 + 12 \times 3$
   (E) $14 + 4 \times 4$

7. $|-8.2 + 6| =$
   (A) −14.2
   (B) −2.2
   (C) 1.8
   (D) 2.2
   (E) 14.2

8. What is the value of $\left\| -4 \right| \times \left(-2 \times |5 - 6| \right) \right\|$?

9. $8 \times -2 - |11 - 14| =$

    (A) −40

    (B) −19

    (C) −13

    (D) 1

    (E) 8

10. What is the additive inverse of $-\frac{2}{3}$?

    (A) $-\frac{3}{2}$

    (B) $-\frac{4}{6}$

    (C) $\frac{2}{3}$

    (D) $\frac{3}{2}$

    (E) $\frac{4}{2}$

11. Which decimal is the multiplicative inverse of $\frac{2}{5}$?

    (A) −0.4

    (B) 0.6

    (C) 1.4

    (D) 1.5

    (E) 2.5

12. Which statement about 0.01 is correct?

    (A) It is the multiplicative inverse of 10.

    (B) It is the multiplicative inverse of 100.

    (C) It is the additive inverse of $-\frac{1}{10}$.

    (D) It is the additive inverse of $\frac{1}{10}$.

    (E) It is the additive inverse of $\frac{1}{100}$.

ANSWER EXPLANATIONS

1. $3 \div 30 = \frac{3}{30}$

To find the value of this expression, follow the first rule of the order of operations. Since each operation is bounded by parentheses, simply evaluate each pair of number within a closed pair of parentheses. Start with $4 + 2$:

$(3 \div (5 \times (4 + 2))) =$

$(3 \div (5 \times 6)) =$

$3 \div 30 = \frac{3}{30}$

This fraction is also equal to $\frac{1}{10}$ and 0.1

2. B

Following the correct order of operations, you must carry out multiplication first. Multiply 8 by -1, and then 3 by 0. Remember the rules for multiplying by -1 and 0: $8 \times -1 = -8$ and $3 \times 0 = 0$, so $8 \times -1 + 3 \times 0 = -8 + 0$.

Zero plus any number is that number. So the value of the expression is -8, and (B) is correct. (A) would come from taking the product of 3 and 0 to be -1 instead of 0. (C) results from taking that product to be 1. (D) results from ignoring the order of operations and performing each one from left to right. (E) would come from combining -1 and 3 as a first step, and then taking the product of 16 and 0 to be 16.

3. B

Perform the operation in parentheses first:

$$-7 - 2 = -9$$

$$8 - 3 \times (-7 - 2) =$$

$$8 - 3 \times -9$$

Now you must carry out the multiplication:

$$3 \times -9 = -27$$

$$8 - 3 \times -9 =$$

$$8 - (-27) = 35$$

(A) is the result of subtracting 3 from 8 instead of multiplying –9 by 3. (D) results from a subtraction error. If you got –5 instead of –9 for the value of –7 – 2, you would go to evaluate 8 – 3 × – 5. (C) would be right if you got 5 as the value of –7 – 2. (E) would be tempting if you overlooked the parentheses in evaluating the expression and multiplied 3 by – 7 as the first step.

4. E

Following the order of operations, handle the expressions inside parentheses first. Set up the subtraction by adding decimal places and lining up the decimals:

$$1.00$$
$$- \underline{0.06}$$
$$0.94$$

$1 - 0.06 = 0.94$
$0.94 \times 7 = 6.58$

(B) is the result of adding one decimal place too many. (D) is the result of a subtraction error. If you subtracted 0.6 instead of 0.06, or if you lined up the numbers incorrectly, you might have gotten 0.4 instead of 0.94 for the difference. 2.8 is the product of 7 and 0.4.

5. E

This question requires you evaluate an expression involving negative numbers. First, divide 15 by –3. Division with negative numbers follows the same basic rules as multiplication.

$$15 \div -3 = -5$$
$$-4 \times (15 \div -3 - 2) = -4 \times (-5 - 2)$$

We must carry out the operations inside parentheses first. So:
$$-4 \times (-5 - 2) = -4 \times -7$$

So the expression has a value of 28, and (E) is correct. (A) would be the result of incorrectly getting a negative product of two negative numbers. (C) results from subtracting 2 from –3 before dividing. You would then divide 15 by –5 to get –3. The product of –4 and –3 is 12.

6. C

Unlike the earlier questions, this one starts by giving you a number, and asks you to find the expression that has that value. Don't be thrown off by this difference; you still need to evaluate expressions. The product of 9 and –9 is –81. Subtracting –81 from a number is like adding 81 to it. The sum of 81 and –9 is 72.

(A) might result from incorrectly multiplying negative numbers. The value of that expression is actually –72. The product of three negative numbers will be negative.

7. D

The absolute value of an expression is the distance of its value from 0 on the number line. Since the question asks for the absolute value, you need to first find the value of –8.2 + 6. Since that value is –2.2, the absolute value of the value is the distance of –2.2 from 0 the number line, which is 2.2.

(E) would result from getting –14.2 instead of –2.2 when adding the numbers. (A) and (B) both fail to account for the absolute value sign. Remember that absolute values are always positive.

8. 8

Here you have absolute values within a larger absolute expression. So you'll have to do some arithmetic before getting the last absolute value. Start with the two absolute value expressions inside the outermost brackets:

$$|-4| = 4$$
$$|5 - 6| = |-1| = 1$$

Substitute these new values in your expression:

$$|4 \times (-2 \times 1)|$$

Now evaluate $4(-2 \times 1)$. Start with numbers within the parentheses:

$$4(-2 \times 1) = 4 \times -2 = -8$$

So $|4(-2 \times 1)| = |-8| = 8$

9. B

This question involves an absolute value expression, but it is only a part of a larger expression. Remember to treat an absolute value expression as though it were an expression in parentheses, so that it has priority in the order of operations.

Since $11 - 14 = -3$, the absolute value of $|11-14|$ is 3.
Therefore, $8 \times -2 - |11-14| = 8 \times -2 -3$

Since $8 \times -2 = -16$,
$8 \times -2 -3 = -16 - 3 = -19$

Keep in mind that absolute value expressions can be subtracted; the absolute value sign *does not* cancel out the minus sign that comes before it. If you did cancel out the minus sign, you might have gotten (C). (A) would result from an incorrect order of operations. (D) and (E) also involve the error of subtracting before multiplying, but the subtraction is done incorrectly in both cases.

10. C

A number and its additive inverse have a sum of 0. Since $-\frac{2}{3}$ is two thirds of a unit to the left of 0 on the number line, you'd have to move two thirds of a unit to the right to get to 0. This means that you must add $\frac{2}{3}$ to $-\frac{2}{3}$ to get a sum of 0. So $\frac{2}{3}$ is the additive inverse of $-\frac{2}{3}$, and (C) is correct. (A) might also be tempting, as it is the multiplicative inverse of $-\frac{2}{3}$. (D) is the multiplicative inverse of the additive inverse.

11. E

You need to convert a fraction to a decimal to answer this question. It would be easiest here to find the inverse of $\frac{2}{5}$ as a fraction, and convert that fraction to a decimal (though you could convert $\frac{2}{5}$ to a decimal and find the inverse of the decimal). The multiplicative inverse of $\frac{2}{5}$ is $\frac{5}{2}$. All we did was switch the numerator and denominator. [To check this, you can find that $\frac{2}{5} \times \frac{5}{2} = \frac{2 \times 5}{5 \times 2} = \frac{10}{10} = 1$]

To convert $\frac{5}{2}$ to a decimal, divide 5 by 2. The result is 2.5. (B) might result from some confusion about the additive inverse, as the sum of 0.6 and $\frac{2}{5}$ is 1. (C) results from taking the product of 0.4 (the decimal value of $\frac{2}{5}$) and 1.4 to be 1. (D) is the result of a division error.

12. B

You can answer this question and rule out incorrect options by finding both the additive and multiplicative inverses of 0.01. Start by converting 0.01 to a fraction. Since the 1 in this decimal is in the hundredths place, the decimal has a value of $\frac{1}{100}$. That should be enough to rule out (E). The additive inverse of $\frac{1}{100}$ is $- \frac{1}{100}$. That rules out (C) and (D); none of the remaining options involve additive inverses. The multiplicative inverse of $\frac{1}{100}$ is $\frac{100}{1}$, or 100.

CHAPTER 2

Algebraic Expressions

WHAT ARE ALGEBRAIC EXPRESSIONS?

In chapter 1 we covered a lot of ground with arithmetic expressions. Now we will go over *algebraic expressions*. Many algebraic expressions are like the kinds we've encountered already, but they also involve something very important: *variables*. A variable is a symbol that stands for a number. They usually take the form of letters.

$$x - 5$$
$$y + 6$$
$$z \times 5$$
$$\frac{a}{5}$$

Each of the letters x, y, z, a is a variable. The value of an algebraic expression depends on the values of the particular variables it contains.

CONCEPTS TO HELP YOU

In order to deal with algebraic expressions, you need to know how to *simplify* and *evaluate* variables. Dealing with the values of variables, when they are available, is very important. It is also important to understand when and how you can perform operations with algebraic terms. We'll talk about three new number properties that will help you in this department.

Variables

A variable is a symbol that stands for an unknown quantity in an expression. While any expression of arithmetic can be evaluated, an algebraic expression can be evaluated only if values for its variables are given.

We can't evaluate the expression $3a + 4a$, for instance, with knowing the value of a (we could simplify it, though).

Even without the value of a in this expression, we know that both instances of a have the same value. If we wanted to write an expression with more than one variable, each having a different value, we would have to use more than one letter.

Values of Terms and Expressions

Many Algebra I questions will contain an algebraic expression and the values of its variables separately. When you are given those values, you can then evaluate the expression. You can evaluate $b - 1$ if you're told what number b equals. It's just a matter of substituting that value into the expression.

So, if $b = 5$, $b - 1 = 5 - 1 = 4$. Likewise, if $x = 4$ and $y = 5$, $x + y = 4 + 5 = 9$.

Combining Terms

Expressions Involving Multiplication

Let's note one thing before going further: though it might not initially appear this way, terms such as $5x$ actually do involve multiplication; $5x$ is the product of 5 and x. In Algebra, we often express products of numbers and variables in this way, and we do without the multiplication sign whenever possible. One reason for this is that since the letter x is used as a variable quite a lot, it gets confusing when you also use a multiplication sign that looks like the letter x. So instead of writing the expression $3 \times y + 5$, we can write $3y + 5$.

When we multiply two numbers together, we will use a new symbol from now on. Instead of writing 7×8, we will write $7 \bullet 8$. This dot is a multiplication sign, and will we use instead of the '$\times$' sign to avoid confusion.

To multiply a term with a variable, such as $3x$, by a number, find the product of the numbers, and put the variables at the end:

$3x \bullet 8 = 24x$

If both terms have variables, both variables wind up in the product:

$3x \bullet 6y = 18xy$

If both terms have the same variable, you need to use *powers* to express the product. We'll get to that in chapter 4.

Combining Like Terms

When it comes to addition and subtraction, you can only combine *like terms*. Two terms are like as long as neither contains a variable that the other doesn't have. So $5x$ and $6x$ are like terms, as are $2y$ and $3y$. For that matter, 2 and 3 are like terms.

None of the following, however, are pairs of like terms:

3 and $3x$

$4x$ and $4y$

$4xy$ and $4x$

If you have a pair of like terms such as $3x$ and $4x$, adding them is just a matter of adding the numbers, and then putting the variable behind the sum. Since $3 + 4 = 7$, then, $3x + 4x = 7x$. Likewise, $6y - 2y = 4y$. Since the expression $7y - 4x$ contains no like terms, it is fully simplified.

More Number Properties

Some number properties are especially helpful when it comes to algebraic expressions.

The Commutative Property

You might find the commutative property to be simplest one: It holds that:

$a + b$ is equal to $b + a$

$a \bullet b$ or ab is equal to $b \bullet a$ or ba

Note that this property holds for addition and multiplication *only*; it doesn't work with subtraction or division. $15 - 8 \neq 8 - 15$ and $10 \div 2 \neq 2 \div 10$.

The Associative Property

The associative property holds that in an expression with exactly two operations, the order of operations doesn't matter if the operations are the same. Let's look at an expression without variables:

$(4 \bullet 3) \bullet 5$

The value of this expression is 60. You can see that the order of operations doesn't matter here. If you moved the parentheses to get the expression $4 \bullet (3 \bullet 5)$, you would also get a value of 60.

We can state this relationship more generally using letters instead of numbers:

$a \bullet (b \bullet c) = (a \bullet b) \bullet c$

$(a \bullet b) \bullet c = a \bullet (b \bullet c)$

So $8 \bullet (6 \bullet b)$ can be rewritten as $(8 \bullet 6) \bullet b$. That expression equals $48 \bullet b$, or $48b$.

In addition, the associative property holds for addition. (Though not for expressions that combine addition or subtraction with multiplication or division):

$a + (b + c) = (a + b) + c$

$(a + b) + c = a + (b + c)$

It does not apply, however for subtraction $[(10 - 8) - 6] \neq [10 - (8 - 6)]$ or division $[(100 \div 20) \div 4] \div [100 \div (20 \div 4)]$.

The Distributive Property

The distributive property holds that:

$a(b + c) = ab + ac$

$ab + ac = a(b + c)$

So $5x + 4x = (5 + 4)x = 9x$.

But let's look at an expression without variables: 4(3 + 5). This is 4 multiplied by 3 + 5. Once again, we are doing without the multiplication sign. When you multiply numbers without a multiplication sign, there must be parentheses between them. Now, the value of 4(3 + 5) is 32 (you should verify that yourself). The distributive property holds that 4(3 + 5) equals 4 • 3 + 4 • 5. If you evaluate that expression, you'll see that it also equals 32.

STEPS YOU NEED TO REMEMBER

The following steps are crucial for the questions we'll encounter soon. Substitution is a step in evaluating expression, while order of operations and number properties are steps in both simplification and evaluation.

Substitution

It is always best to simplify an expression (when possible) before carrying out substitutions. If you simplify before you substitute and evaluate, you'll likely have fewer overall steps than if you performed all substitutions first.

If $x = 2$, you should evaluate $3x - 2x$ by simplifying first: $3x - 2x = x$. So the value of the expression is 2. If you substitute 2 for x first, however, you get:

$$= 3(2) - 2(2)$$
$$= 6 - 4$$
$$= 2$$

This process involves more steps. Naturally, the more steps you have to deal with, the greater your chances of making a mistake.

Once you get to the point where your algebraic expression cannot be further simplified, you can go ahead and carry out substitutions. When you're dealing with more that one variable, be sure to substitute the right value for the right variable.

Order of Operations

Once you have finished all your substitutions, you must still follow the order of operations to evaluate the expression. The order also holds for simplifying algebraic expressions.

Applying the Number Properties

Using the Associative Property

Suppose you need to simplify $3a + (4a + 5)$. You would like to combine the like terms—$3a$ and $4a$—but they are separated by a parenthesis sign. The associative property comes in handy here. We can use it to rewrite our expression: $(3a + 4a) + 5$.

Now that the like terms are together within a pair of parentheses, we can add them to get the simplified expression $7a + 5$.

Using the Commutative Property

Since the commutative property allows us to reverse the two terms in any addition or multiplication operation, it can be helpful in simplifying expressions.

To combine $2x$ and $3x$ in $2x + 5 + 3x$, we have to use the commutative property to get the expression $2x + 3x + 5$.

Distributing Minus and Negative Signs

The expression $-(4 + x)$ equals $-4 - x$.

How do we know this? Any negative number is equal to its additive inverse multiplied by -1. So

$-(4 + x) = -1 \cdot (4 + x)$.

Now we can apply the distributive property, such that $-1 \cdot (4 + x) = -1 \cdot 4 + -1 \cdot x = -4 - x$. So if you are subtracting a term like $4 + x$, the minus sign would distribute, and you subtract both 4 and x

STEP–BY–STEP ILLUSTRATION OF THE 5 MOST COMMON QUESTION TYPES

Now let's tackle some common question types. There are really only two basic question types when it comes to algebraic expressions: simplifying and evaluating.

Question 1: Simplifying Algebraic Expressions—Addition

$5x + 4x + 11 =$

(A) $4x + 16x$

(B) $5x + 15$

(C) $9x + 11$

(D) $20x$

(E) $20x + 9$

You're being asked to *simplify* an expression. To simplify is to rewrite it using the fewest possible terms. You can do this by combining terms where possible.

Terms with the same variables such as $5x$ and $4x$ can be added. The sum of those terms is $9x$. You keep the variable and add the numbers that come before it. You can do this, however, *only if the variables are the same*. So you cannot combine $5x$ and $4y$ in this way.

So we are able to combine $5x$ and $4x$, and our expression is now $9x + 11$. Can we combine these two terms? The answer is *no*. You cannot add a number to a term with a variable. You can only add or subtract *like terms*. You have a pair of like terms as long as neither of them contains a variable that the other doesn't have. *This* means that the expression is fully simplified, and so **(C) is correct**.

Choice (D) is the incorrect result of combining 11 with $9x$. You could get $20x$ by adding $11x$ to $9x$. (A) and (B) are also the result of combined *unlike terms*.

The rule about combining *like terms* applies to subtraction as well. You will find that is does not apply to multiplication and division.

Question 2: Simplifying Expressions – Multiplication and Division

Which of the following is a simplified form of $8(6b) - 20b \div 4$?

(A) $-28b$

(B) 7

(C) $7b$

(D) $43b$

(E) $48b - 5$

Start by multiplying 8 by $6b$. To multiply a term with a variable by a number, find the product of the two numbers (8 and 6 here). Put the variable at the end, and you have you product. So $8 \bullet 6b = 48b$.

Now carry out the division operation. $20b$ divided by 4 is $5b$; you can handle division just as you would multiplication. So the expression we started with has now been simplified to $48b - 5b$. You can simplify this even further by carrying out the subtraction. Using the approach we went over in question 1, you should get $43b$ as the result $[48b - 5b = (48 - 5)b = 43b]$.

(D) is the correct answer. Choice (A) results from subtracting $20b$ from $6b$ first (remember the order of operations). (C) results from performing division last. (E) results from getting 5 instead of $5b$ when dividing $20b$ by 4.

Question 3: Simplifying Expressions With More Than One Variable

$5xy - 3x(2 + y) + 4(x + y) =$

(A) $2xy - 10x + 4y$

(B) $2xy - 2x + 4y$

(C) $5xy - 10x + 7y$

(D) $5xy - 2x + 7y$

(E) $11xy - 6x$

You have multiple variables here, so you'll have to be careful about combining like terms. Though there are two operations found within parentheses, there's nothing you can do to carry those out. Neither $2 + y$ nor $x + y$ can be further simplified.

There are other multiplication operations to carry out here, though. You can multiply $3x$ by $2 + y$, and you can multiply 4 by $x + y$. This is possible because of the distributive property. Carry out the multiplication to simplify both parts of the expression:

$3x \bullet (2 + y) = 3x \bullet 2 + 3x \bullet y = 6x + 3xy$

$4 \bullet (x + y) = 4x + 4y$

So we can rewrite our original expression as

$5xy - (6x + 3xy) + 4x + 4y$

Notice that the expression $6x + 3xy$ is still in parentheses. This is because the expression $3x(2 + y)$, from which we got $6x + 3xy$, is being subtracted. We need the parentheses to show that $6x + 3xy$, and not just $6x$, is being subtracted.

So how do you subtract an expression with more than one term? The distributive property comes into play again. As we explained earlier, the negative sign *distributes* inside the parentheses, so that both terms get subtracted:

$5xy - (6x + 3xy) =$

$5xy - 6x - 3xy =$

$5xy - 6x - 3xy + 4x + 4y$

This can be simplified further still by combining like terms. (Remember that you have a pair of like terms as long as neither term contains a variable that the other doesn't have. $5xy$ and $3xy$ are like terms, but $4x$ and $4y$ are not).

What we would like to do first is combine $5xy$ and $3xy$. We need to get them next to each other. It would be great to swap the $6x$ and $3xy$. To accomplish this, however, we need to back up a step.

$5xy - (6x + 3xy) + 4x + 4y$ Now use the commutative property

$5xy - (3xy + 6x) + 4x + 4y$

Why did we have to back up? Since the commutative property only works with addition and multiplication, we couldn't use it once we distributed the minus sign. You'll find sometimes that once you see how to get to a solution, you'll need to back up a little to make things work.

Now we can distribute the minus sign:

$5xy - 3xy - 6x + 4x + 4y$

Now we have an expression with just addition and subtraction operations to carry out. We start with the first one on the left, and subtract $3xy$ from $5xy$ to get $2xy$, leaving us with:

$2xy - 6x + 4x + 4y$

We can't combine $2xy$ and $6x$, so we move to the next operation. *Be careful here*. We do not add $4x$ and $6x$. It helps approach this by rewriting the expression:

$2xy + (-6x) + 4x + 4y$

Rather than subtracting a term, you are *adding the negative of that term*. If you put it that way, you can add $-6x$ and $4x$ to get $-2x$.

$2xy + - 2x + 4y$

The first plus sign doesn't need to be there, so we can write

$2xy - 2x + 4y$

This expression cannot be simplified any further, so **(B) is correct.** (A) results from adding $6x$ and $4x$ instead of $-6x$ and $4x$. (E) results from an incorrect use of the distributive property (getting $4xy$ instead of $4x + 4y$ from $4(x + y)$).

Question 4: Evaluating Algebraic Expressions With One Variable

If $y = -5$, then $6y + 4 - 2y =$

(A) -44

(B) -40

(C) -36

(D) -24

(E) -16

In the first three questions in this chapter, you could not evaluate the expressions you were given because they contained variables with unknown values. Here, things are different. The question gives you the value of y.

When the value of a variable is specified, the expression can be evaluated *by substituting the value for each instance of the variable*; wherever you see y, just replace it with -5. You will then have an arithmetic expression to evaluate.

It is usually easier to evaluate algebraic expressions by simplifying them before substituting the value of the variable. In this question, you can simplify $6y + 4 - 2y$ before putting in the value of y. There are two like terms here ($6y$ and $2y$), and you can get them together with the help of the commutative property.

Let's switch the first two terms:

$4 + 6y - 2y$ Now combine the terms with variables

$4 + 4y$

By simplifying as a first step, we have set things up so that we need only substitute -5 for y once. When we make the substitution, we get:

$4 + 4 \bullet -5$ Carry out the multiplication first

$4 + -20$

$4 - 20 = -16$

(E) is the answer. (C) is what you might get by substituting before simplifying, if you got 10 instead of -10 for the value of $2y$.

Question 5: Evaluating Algebraic Expressions With More Than One Variable

If $h = 7$ and $k = -9$, $6k - 5h - hk =$

This last question is open-ended. The expression involves two variables, and the question specifies both of their values. When dealing with multiple variables, be extra careful when performing your substitutions.

This question is actually written in a way that is meant to trick you: the order in which variables appear in the expression and the order in which their values are specified are different. Many incorrect answers to this type of question are the result of a wrong substitution. It always helps to substitute one value at a time, and then simplifying after each substitution, when possible.

Start with the variable h. Since $h = 7$, we can substitute that number for the first instance of h to get

$6k - 5(7) - hk$

$6k - 35 - hk$ Now substitute for the other instance of h

$6k - 35 - 7k$ Substitute for the first k

$6(-9) - 35 - hk$

$-54 - 35 - 7k$

$-89 - 7k$ Substitute the last variable for its value of -9

$-89 - 7(-9)$

$-89 - (-63)$

$-89 + 63$

So -26 is the correct answer. This might seem like a lot of steps to perform, but you'll move through them very quickly once you've had some practice.

CHAPTER QUIZ

1. $4a(2b + 3c) =$
   (A) $6ab + 7ac$
   (B) $8ab + 3c$
   (C) $8ab + 7ac$
   (D) $8ab + 12ac$
   (E) $20abc$

2. Simplify $4x + (5y + 6x)$.
   (A) $4x + 11y$
   (B) $10x + 5y$
   (C) $15x$
   (D) $15xy$
   (E) $15x + 5y$

3. Which expression can be further simplified?
   (A) $3x - 4$
   (B) $4y + 11y$
   (C) $5x + 5y$
   (D) $6y + 6$
   (E) $7z + xz$

4. $5c - 2c(2d + 5) =$
   (A) $-5c - 4cd$
   (B) $-5c + 4cd$
   (C) $5 + 6cd$
   (D) $15c - 4cd$
   (E) $15c + 6cd$

5. If $g = -3$, which expression has a value of 12?
   (A) $(3g + 7g) \bullet -1$
   (B) $3g + 7g \bullet -1$
   (C) $3g \bullet -1 - 7g$
   (D) $7g \bullet -1 - 3g$
   (E) $7g \bullet (-1 - 3g)$

6. If $n = 13$, $2n + 5(3 + n) =$

7. If $s = 0$ and $t = 8$, then $7 + 2t - 4s =$
   (A) -25
   (B) 0
   (C) 19
   (D) 23
   (E) 72

8. If $a = -7$ and $b = 9$, what is the value $|2a - b| - 2|ab|$
   (A) -159
   (B) -103
   (C) -99
   (D) 103
   (E) 159

ANSWER EXPLANATIONS

1. D

Use the distributive property here. When you multiply a term such as $4a$ by an expression such as $2b + 3c$, you multiply $4a$ by each term, and add the products.

$$4a(2b + 3c) = (4a \bullet 2b) + (4a \bullet 3c) = 8ab + 12ac$$

(B) is the result of multiplying $4a$ by $2b$ without also multiplying $4a$ by $2b$. (A) and (C) both involve incorrect addition of terms that should instead be multiplied. (E) involves an attempt to simplify the expression a step further than is possible, by adding $8ab$ and $12ac$ in an incorrect way.

2. B

Since $4x$ and $6x$ are like terms, you'll want to reorganize the expression so they can be combined. Use the commutative property on $4x + (5y + 6x)$ to get $4x + (6x + 5y)$. Then use the associative property to get $(4x + 6x) + 5y$. Now you add the like terms to get $10x + 5y$.

(A) results from combining unlike terms. But $5y + 6x$ cannot be simplified. (C) and (D) result from combining all three terms. And with (D), note that the only to get an "xy term" from a "x term" and a "y term" is by multiplication.

3. B

$4y + 11y$ is the only expression containing like terms. (C) and (D) might be tempting because the terms in each of those have a common number. (E) might be tempting as well because the terms in that expression have a variable in common.

4. A

In this expression, the part inside the parentheses cannot be simplified, so you should move on to the multiplication. Since you're multiplying $2c$ by $2d + 5$, you'll use the distributive property:

$$2c(2d + 5) - 2c \bullet 2d + 2c \bullet 5$$

$$4cd + 10c$$

$$5c - 2c(2d + 5) = 5c - (4cd + 10c)$$

Now look to combine the like terms, $5c$ and $10c$, but you'll need to get them closer. Use the commutative property to switch $4cd$ and $10c$:

$5c - (10c + 4cd)$

It's not just $10c$ that is being subtracted here. You need to subtract the entire expression $10c + 4cd$. When you do that, the subtraction sign "distributes," and you get $5c - 10c - 4cd$. After combining the like terms, you get $-5c - 4cd$.

(A) is the answer. (B) is the result of forgetting to distribute the negative sign. (E) results from subtracting $2c$ from $5c$ before multiplying by $2d + 5$. (D) also involves an incorrect distribution that gets $5c + 4cd - 10c$ instead of $5c - 10c - 4cd$.

5.   B

With a question like this, you'll likely need to evaluate the expressions one at a time, until you find the one that has a value of 12. Start with (A):

$$(3g + 7g) \bullet -1 =$$
$$(3(-3) + 7g) \bullet -1 =$$
$$(-9 + 7g) \bullet -1 =$$
$$(-9 + 7(-3) \bullet -1 =$$
$$(-9 + (-21)) \bullet -1 =$$
$$(-9 - 21) \bullet -1 =$$
$$-30 \bullet -1 =$$
$$30$$

So (A) can be ruled out. Let's take a shortcut; now that we've worked through one expression, we've found that $3g = -9$ and $7g = -21$. Why not substitute those values instead? With choice (B), we get:

$$3g + 7g \bullet -1 =$$
$$-9 + -21 \bullet -1 =$$
$$-9 + 21 =$$
$$12$$

So (B) is correct. You can confirm the values of the remaining choices:

$$3g \bullet -1 - 7g = 30$$
$$7g \bullet -1 - 3g = 30$$
$$7g \bullet (-1 - 3g) = -168$$

6. 106

Though this expression seems to require the use of the distributive, it really doesn't. You can first substitute for the variable inside the parentheses, and then carry out the addition inside. We know that $n = 13$.

$$2n + 5(3 + n) =$$
$$2n + 5(3 + 13) =$$
$$2n + 5(16) =$$
$$2n + 80$$

It would have taken longer to get to this point if you had used the distributive property first.

Now, replace n with 13 again, and simplify:

$$2(3) + 80 =$$
$$26 + 80 =$$
$$106$$

7. D

Substitute the values of the variables into the expression, one at a time. It's best to simplify as much as possible before each substitution. Since we don't have any like terms to start with, though, there's nothing we can do before our first substitution. So replace s with 0 and simplify the result:

$$7 + 2t - 4s =$$
$$7 + 2t - 4(0) =$$
$$7 + 2t - 0 =$$
$$7 + 2t$$

Remember that the product of 0 and any number is 0. Now replace t with 8 and simplify:

$$7 + 2t =$$
$$7 + 2(8) =$$
$$7 + 16 =$$
$$23$$

(D) is correct. (A) is the result of mistakenly substituting 0 for t and 8 for s. (D) results from taking the product of 4 and 0 to be 4 instead of 0.

8.   B

You are asked here to evaluate an algebraic expression with absolute values. Since absolute value expressions should be treated like operations in parentheses, you should evaluate them first.

$$|2a - b|$$
$$= |2(-7) - 9|$$
$$= |-14 - 9|$$
$$= |-23|$$
$$= 23$$

So:

$$|2a - b| - 2|ab| = 23 - 2|ab|$$

$$|ab|$$
$$= |(-7)(9)|$$
$$= |-63|$$
$$= 63$$

So $2|ab| = 2(63) = 126$

$$|2a - b| - 2|ab| = 23 - 126 = -103$$

(B) is correct. (A) involves an arithmetic error: taking 23 – 126 to be the sum of 23 and 126 with a negative sign added. (C) results from substituting 9 for *a* and –7 for *b*, rather than the reverse. (D) might be tempting if you thought the absolute value sign here suggested that the value of the *whole* expression should be an absolute value.

Factors

WHAT ARE FACTORS?

Every number is the product of at least one pair of numbers. *Factors* are those numbers that are multiplied. Since 30 is the product of 5 and 6, 5 is a factor of 30, and so is 6. Similarly, 7 is a factor of 28 because that number is the product of 7 and 4.

Factors play several important roles in Algebra I. Factoring, which is the process of identifying a number's factors, is important for dealing with fractions. It is also important for simplifying different kinds of expressions.

CONCEPTS TO HELP YOU

In addition to factors, we'll review here several related concepts, including prime numbers and multiples, all of which help us to deal with fractions. The *greatest common factor* and *least common multiple* allow us to add, subtract, and simplify them.

Factors

Another way to define a *factor* of a given number is this: For any given number, if you can divide it by one integer to get another integer, then that integer is a factor of the number, as long as there is no remainder after dividing. Since the result of dividing 10 by 6 is 1 with a remainder of 4, 6 is not a factor of 10. On the other hand, since 10 divided by 5 is exactly 2, 5 is a factor of 10. For that matter, so is 2.

USE FACTORS TO DEFINE ODD AND EVEN NUMBERS

An even number is one that has 2 as a factor. An odd number is a number that does not have 2 as a factor.

Negative numbers can also be factors: –15 is a factor of 60, because 60 ÷ –4 = –15. Negative factors play a significant role in polynomials (chapter 10) and quadratic equations (chapter 11).

We'll also treat variables as factors of algebraic terms. For example, 6 and x are both factors of $6x$. For that matter, $3x$ is a factor of $6x$, since $6x \div 2 = 3x$.

Prime Numbers and Prime Factors

A *prime number* is a number that has exactly two factors: 1 and itself. A number that is not prime (and has more than two factors) is called a *composite number*.

PRIME NUMBERS ARE USUALLY ODD

Most prime numbers are odd. Even numbers always have 2 as a factor, so the only even prime number is 2.

There are prime numbers so large that even powerful computers takes hours to identify them, but chances are that you'll need only to know a few to succeed in Algebra I. The prime numbers smaller than 100 are:

2, 3, 5, 7, 11, 13, 17, 19, 23, 29, 31, 37, 41, 43, 47, 53, 59, 61, 67, 71, 73, 79, 83, 89, 97

Every odd and even integer is *composed* of *prime factors*. By this we mean that every number can be expressed as a product of prime numbers. Some examples:

$$15 = 3 \bullet 5$$
$$36 = 2 \bullet 2 \bullet 3 \bullet 3$$
$$52 = 2 \bullet 2 \bullet 13$$

Greatest Common Factors

The greatest common factor (GCF) of two numbers is the largest number that is a factor of both. To identify the GCF of a number, it helps to be able to find the prime factors.

Let's find the GCF of 6 and 8. With numbers so small, it's not hard to just list the factors of both and then pick out the largest one that appears on both lists:

Prime factors of 6:   1, 2, 3, 6

Prime factors of 8:   1, 2, 4, 8

Since 2 is the largest number that appears on both list, 2 is the GCF of 6 and 8.

When it comes to larger numbers, there's a quicker way to find the GCF. It involves listing the prime factors of each number. Take 24 and 36:

$24 = 2 \bullet 2 \bullet 2 \bullet 3$

$36 = 2 \bullet 2 \bullet 3 \bullet 3$

Now identify each time a prime factor appears in *both* products. Since 2 shows up 3 times as a factor of 24 and 2 times as a factor of 36, we say that 2 appears twice in both products. Since 3 shows up 1 time as a factor of 24 and 2 times as a factor of 36, we say that 3 appears once in both products. So we'll list the prime factors of 24, underlining the prime factors that also appear in 36:

24: 2, 2, 2, 3

We'll call these the common prime factors of both numbers. The GCF of 24 and 36 is the product of these common prime factors. The product of 2, 2, and 3 is 12, and so 12 is the GCF.

NEGATIVE FACTORS

When finding the GCF, don't worry about negative factors, since negative numbers are always smaller than positive ones.

We'll talk about finding prime factors shortly. Once you master that, you'll be able to find the GCF of any group of numbers.

Least Common Multiples

The *least common multiple* (LCM) of two numbers is another important concept. A *multiple* of a number is a product of that number and any integer. Since 16 is the product of 8 and 2, and 24 is the product of 8 and 3, 16 and 24 are both multiples of 8. (A multiple is the opposite of a factor, in a sense. Take any two numbers; if the first is a factor of the second, then the second is a multiple of the first. Since 3 is a factor of 9, 9 is a multiple of 3.)

The LCM of two numbers is the smallest number that is a multiple of both. The LCM of 2 and 3 is 6. The LCM of two numbers is not always their product; it can be a smaller number. When the LCM is smaller than the product, you can use prime factors to find it.

Let's find the LCM of 6 and 8. It's not hard to figure out that 48 is a common multiple of those numbers, but is it their *least* common multiple?

    Prime factors of 6:   2, 3

    Prime factors of 8:   2, 2, 2

To find the LCM of 6 and 8, get the product of the prime factors of both numbers. This gets tricky because each prime factor that appears in both numbers is factored in only once. Look at the list of factors of each number again:

    6:   2, 3

    8:   2, 2, 2

Since 2 is a prime factor of 6, you don't count the first prime factor of 2 in 8. Just cross it out, and underline the 2 that is a prime factor of 6:

    6:   <u>2</u>, 3

    8:   ~~2~~, 2, 2

You keep the remaining prime factors of 8. The idea is that each instance of a prime factor appearing in both numbers gets counted only once. We underlined the first 2 so you would have a reminder that it doesn't cancel out any of the remaining 2's.

Now you find the product of these remaining prime factors:

$2 \bullet 3 \bullet 2 \bullet 2 = 24$

So 24 is the LCM of 6 and 8.

Let's take another example, finding the LCM of 18 and 24 by highlighting unique prime factors.

18: 2, 3, 3

24: 2, 2, 2, 3

Since 2 appears once as a factor of 18, we crossed it out once under 24. Since 3 appears once as a factor of 18, we crossed it out once under 24. Now we find the product of the remaining numbers.

$2 \bullet 3 \bullet 3 \bullet 2 \bullet 2 = 72$

So 72 is the LCM of 18 and 24

LCM OF TWO NUMBERS

Every number is a multiple of itself, because every number is a product of itself and 1. Keeping this in mind, you may find that the LCM of two numbers is actually one of them. The LCM of 2 and 10, for instance, is 10.

Simplified Fractions and Common Denominators

We can apply the concept of the GCF to simplifying fractions. You can simplify a fraction by rewriting it as one of equal value with the smallest possible numerator and denominator.

Take the fraction $\frac{1}{2}$. We know that $\frac{1}{2} \bullet 1 = \frac{1}{2}$, and since the number 1 can be rewritten as a fraction, then $\frac{1}{2} \bullet \frac{1}{1} = \frac{1}{2}$. And since every number divided by itself equals 1, we can express the value of 1 as any fraction where the numerator and denominator are equal.

So $1 = \frac{1}{1} = \frac{2}{2} = \frac{3}{3} = \frac{5}{5} = \frac{20}{20}$, and so on.

But what if we multiply $\frac{1}{2}$ by one of those fractions?

$$\frac{1}{2} \bullet \frac{3}{3} = \frac{3}{6}$$ [Remember the rules for multiplying fractions]

Since $\frac{3}{3} = 1$, it follows that $\frac{1}{2} = \frac{3}{6}$. We can say, moreover, that $\frac{1}{2}$ is a simplified form of $\frac{3}{6}$; the fractions are equal, but $\frac{1}{2}$ is the form with the small possible fraction equal with that value (as $\frac{3}{6}$ also has a value of $\frac{1}{2}$).

You can simplify a fraction by dividing it by a fraction equal to 1. Take $\frac{3}{6}$, and divide it by $\frac{3}{3}$:

$$\frac{3}{6} \div \frac{3}{3} = \frac{3 \div 3}{6 \div 3} = \frac{1}{2}$$

DIVISION OF FRACTIONS

We usually handle the division of fractions using the multiplicative inverse. Here, we're using it differently, to explain our point.

To reduce a fraction to its simplified form, *divide the numerator and the denominator by their GCF*. This insures that the result is equal to the original fraction, with the smallest possible numerator and denominator.

Take the fraction $\frac{25}{35}$. The GCF of 25 and 35 is 5. If we divide both the numerator and denominator by 5, the result is $\frac{5}{7}$. That fraction is a simplified form of $\frac{25}{35}$.

Being able to simplify fractions is important on many multiple-choice tests. The correct choice might be a simplified form of the solution you have reached. Without simplifying, you wouldn't be able to recognize the correct choice.

The LCM comes in handy for another purpose: finding the common denominator of two fractions. Since two fractions can be combined with addition or subtraction only if the denominators are the same, you would be stuck evaluating $\frac{3}{4} + \frac{5}{6}$. You could multiply each fraction by a fraction equal to 1 in order to get fractions with equal denominators. The denominator value you are looking for is the LCM of the denominator you have.

Since the LCM of 4 and 6 is 12, you can rewrite $\frac{3}{4}$ and $\frac{5}{6}$ as fractions with denominators of 12.

Since 12 is the product of 4 and 3, you have to multiply $\frac{3}{4}$ by $\frac{3}{3}$ to get an equal fraction with a denominator of 12. The result of that is $\frac{9}{12}$.

Since 12 is the product of 6 and 2, you have to multiply $\frac{5}{6}$ by $\frac{2}{2}$ to get an equal fraction with a denominator of 12. The result of that is $\frac{10}{12}$.

$$\frac{3}{4} + \frac{5}{6} = \frac{9}{12} + \frac{10}{12} = \frac{19}{12}$$

STEPS YOU NEED TO REMEMBER

The steps we'll go over here will help you to find GCF's and LCM's. Though we've reviewed the concepts, it's important that you are clear on how to find a number's prime factors, and that you have factored numbers completely and correctly.

Breaking Down Numbers into Prime Factors

Your first step in finding a GCF or LCM is to break down your number into its prime factors. Start with the lowest prime number (2 if the number is even, or 3 if it is odd) and check whether it is a factor. Let's take the number 90:

Since 90 is even, it can be divided by 2. The result is 45, so you can rewrite 90 as 2 • 45. Now 45 can't be divided by 2, so go on to check the next prime number, 3. Since 45 divided by 3 is 15, rewrite your expression as 2 • 3 • 15.

So far, you have identified 2, 3, and 15 as factors of 90. But don't move on the next prime number just yet; it's possible that the last factor you got can also be divided by 3. In fact, 15 divided by 3 is 5, and so you can write 2 • 3 • 3 • 5.

Since 5 is a prime number, we're done. We have found all of the prime factors of 90.

Let's try one more example and factor 84:

$$84 =$$
$$2 • 42 =$$
$$2 • 2 • 21 =$$
$$2 • 2 • 3 • 7$$

You keep dividing the last number in your expression by the lowest prime number that works.

Checking Factors and Multiples

Once you think you've identified a GCF or LCM, you should check your work. Once you find that 16 is the GCF of 32 and 80, for example, evaluate 32 ÷ 16 and 80 ÷ 16. Since the values are 2 and 5, respectively, you would have confirmed that 32 and 80 are multiples of 16. There should never be a remainder when you carry out these divisions.

With the LCM, divide the LCM by each number to confirm that they are true factors of it. If either division operation involves a remainder, you made a mistake. Another thing you might do to be sure that you've found the LCM is to multiply the two numbers. The LCM might be smaller than these numbers, but it cannot be larger. If it is, you made a mistake along the way.

STEP–BY–STEP ILLUSTRATION OF 5 MOST COMMON QUESTION TYPES

Now we're ready to examine the common questions types you're likely to encounter.

Question 1: Identifying Prime Numbers

Which of the following factors of 468 is a prime number?

(A) 4

(B) 9

(C) 12

(D) 13

(E) 27

This question doesn't actually require you to factor 468; you need only identify the prime number among the choices. Since a prime number has no factors other than 1 and itself, we can rule out (A) and (C); as even numbers, they each have 2 as a factor. Since 9 is divisible by 3, (B) is also out.

However, 13 is a prime number, and so **(D) is correct.** You would find that the only number less than 13 that divides into it evenly is 1. Choice (E) is incorrect, as 3 and 9 are factors of 27. Odd numbers are always tempting as multiple-choice options; it may be harder to see that they are composite numbers when 2 is not one of their factors.

Question 2: Factoring

Which two numbers are both factors of 780?

(A) −5 and −18

(B) 7 and −11

(C) 12 and −13

(D) 15 and 16

(E) 20 and 24

This question requires a process of elimination; rather than trying to create a list of all of the factors of 780 and comparing it to the answer choices, test the answer choices by dividing. Remember that a pair of factors of a number need not have that number as a product. In fact, you would find that none of the pairs in the answer options have products of 780.

Remember, too, that a negative number can be a factor of a positive number. If you forgot that, you would be tempted to rule out (A), (B), and (C). Since 780 is divisible by -13, that negative number is a factor. 780 also happens to be divisible by 12, and so **(C) is correct**. (D) is tempting because 15 is a factor of 780 ($780 \div 16 = 48.75$ [or 48 with a remainder of 12]), but 16 is not.

Question 3: Greatest Common Factors

What is the greatest common factor of $120x$ and $135y$?

(A) 5

(B) 15

(C) $15x$

(D) 45

(E) $45xy$

Here you need to find the GCF of two algebraic terms. Find the product of the common prime factors, breaking down each term:

$$120x = x \bullet 120$$
$$= x \bullet 2 \bullet 60$$
$$= x \bullet 2 \bullet 2 \bullet 30$$
$$= x \bullet 2 \bullet 2 \bullet 2 \bullet 15$$
$$= x \bullet 2 \bullet 2 \bullet 2 \bullet 3 \bullet 5$$

$$135y = y \bullet 135$$
$$= y \bullet 3 \bullet 45$$
$$= y \bullet 3 \bullet 3 \bullet 15$$
$$= y \bullet 3 \bullet 3 \bullet 3 \bullet 5$$

Now we can list the prime factors of each term, highlighting each instance where a prime factor appears on both lists.

120x: x, 2, 2, 2, <u>3</u>, <u>5</u>

135y: y, 3, 3, <u>3</u>, <u>5</u>

So 3 and 5 each appear once in both products. This means that the GCF of 120x and 135y is the product of 3 and 5, which is 15. **(B) is the answer.**

(C) and (E) contain variables; since neither variable is a factor of both 120x and 135y, they cannot be part of the GCF. (A) might be tempting because it is easy to recognize as a common factor of 120x and 135y. Since 15 is greater, however, 5 cannot be the GCF.

Question 4: Least Common Multiple

What is the least common multiple of 16x and 12y?

(A) 24y

(B) 32x

(C) 48xy

(D) 96x

(E) 192xy

This question can be handled with prime factoring as well, but once you have the prime factors of the terms, they're handed differently than for GCF. Let's break down each term into prime factors:

$$16x = x \bullet 16$$
$$= x \bullet 2 \bullet 8$$
$$= x \bullet 2 \bullet 2 \bullet 4$$
$$= x \bullet 2 \bullet 2 \bullet 2 \bullet 2$$

$$12y = y \bullet 12$$
$$= y \bullet 2 \bullet 6$$
$$= y \bullet 2 \bullet 2 \bullet 3$$

And now, list the prime factors of each term. Each time a prime factor appears on list, we can cross off one instance of the same factor on the other list. We'll cross out both prime factors of 2 on the 12y list, because 2 appears at least that many times on the 16x list.

16x: x, 2, 2, 2, 2

12y: y, ~~2~~, ~~2~~, 3

The LCM is the product of all of the remaining factors.

$x \bullet 2 \bullet 2 \bullet 2 \bullet 2 \bullet y \bullet 3 = 48xy$

So (C) is correct. (E) might be tempting because $192xy$ is a common multiple of the two terms. In light of choice (C), however, it is not the LCM.

Question 5: Adding and Simplifying Fractions

What is the value of $\frac{3}{6} + \frac{1}{10}$?

(A) $\frac{1}{4}$

(B) $\frac{2}{5}$

(C) $\frac{8}{15}$

(D) $\frac{3}{5}$

(E) $\frac{19}{30}$

In order to combine $\frac{3}{6}$ and $\frac{1}{10}$, you need to convert the fractions so that they have common denominators. This common denominator can be the LCM of the denominator of the fractions. Since the LCM of 6 and 10 is 30, we want to convert each fraction to one with a denominator of 30.

You have to multiply 6 by 5 to get 30, so you have to multiply $\frac{3}{6}$ by $\frac{5}{5}$. The result is $\frac{15}{30}$. And you have to multiply 10 by 3 to get 30, so you have to multiply $\frac{1}{10}$ by $\frac{3}{3}$, for a product of $\frac{3}{30}$.

$$\frac{3}{6} + \frac{1}{10} = \frac{15}{30} + \frac{3}{30} = \frac{18}{30}$$

You'll notice that $\frac{18}{30}$ is not one of the answer choices, so we must simplify. Divide the numerator and the denominator by their GCF. The GCF of 18 and 30 is 6.

$$\frac{18 \div 6}{30 \div 6} = \frac{3}{5}$$

(D) is the answer. (A) is the result of simply adding the numerators and the denominators to get $\frac{4}{16}$, which equals $\frac{1}{4}$. (B) would result from getting the correct sum of $\frac{18}{30}$ but then taking 2 instead of 3 to be the result of dividing 18 by 6.

CHAPTER QUIZ

1. A prime factor of 56 is
   (A) 3
   (B) 4
   (C) 5
   (D) 6
   (E) 7

2. A complete list of the prime factors of 90 is
   (A) 2, 2, 2, 5
   (B) 2, 2, 3, 5
   (C) 2, 3, 3, 3
   (D) 2, 3, 3, 5
   (E) 2, 3, 5, 5

3. What is the greatest common factor of $72xy$ and $84y$?

   (A) $6y$

   (B) $12y$

   (C) $12xy$

   (D) $18xy$

   (E) $24y$

4. $90x$ is the least common multiple of which pair of terms?

   (A) $5x$ and 9

   (B) $18x$ and 2

   (C) $15x$ and 18

   (D) $30x$ and $3x$

   (E) 45 and 2

5. Simplify $\frac{20a}{8}$.

   (A) $\frac{5a}{2}$

   (B) $\frac{3a}{2}$

   (C) $\frac{3}{2}$

   (D) $2a$

   (E) $3a$

6. If $r = 6$ and $s = 4$, what is the value of $\frac{2s}{3r}$?

   (A) $\frac{1}{4}$

   (B) $\frac{4}{9}$

   (C) $\frac{1}{2}$

   (D) $\frac{2}{3}$

   (E) 1

7. $\frac{25}{3}$ is a simplified form of

   (A) $\frac{375}{75}$

   (B) $\frac{300}{36}$

   (C) $\frac{625}{60}$

   (D) $\frac{500}{45}$

   (E) $\frac{600}{48}$

8. $\frac{2x}{6} + \frac{3}{8} =$

   (A) $\frac{2x + 3}{8}$

   (B) $\frac{x + 2}{4}$

   (C) $\frac{8x + 3}{24}$

   (D) $\frac{8x + 9}{24}$

   (E) $\frac{2x + 3}{6}$

ANSWER EXPLANATIONS

1. E

"(A) and (C) are prime numbers, but neither is a factor of 56. (B) is a factor of 56, but it is not prime. (D) is not a prime number.

2. D

Follow the steps for factoring a number into primes:

$$90 = 2 \bullet 45$$
$$= 2 \bullet 3 \bullet 15$$
$$= 2 \bullet 3 \bullet 5 \bullet 5$$

(C) might result from taking 3, 3, and 3 instead of 3, 3, and 5 to be the prime factors of 45. (E) might result from taking 5 and 5 instead of 3 and 5 to be the prime factors of 15.

3. B

The GCF is the product of the all of the prime factors the expressions have in common. The factors are listed below, with the common ones underlined.

$72xy$: 2, 2, 2, 3, 3, x, y
$84y$: 2, 2, 3, 7, y, y

The common prime factors are 2, 2, 3, and y. The product of those factors is $12y$, so (B) is correct. (A) would be the result of counting each unique prime factor once.

4. C

Examine each choice to find the pair of terms that have a LCM of $90x$. The LCM of two terms cannot be greater than their product. Since the product of $5x$ and 9 is $45x$, (A) is out. Same thing for (B) and (E). In (D), you might notice that $30x$ is a multiple of $3x$. This means that $30x$ is the LCM of both of those terms.

This leaves (C). Factoring each term gives us these lists of prime factors:

$15x$: x, 3, 5
18: 2, 3, 3

Since 3 appears once on both lists, we can cross out it out once on the list of prime factors of 18. The LCM is the product of the remaining prime factors: $x \bullet 3 \bullet 5 \bullet 2 \bullet 3 = 90x$.

5. A

To simplify this fraction, find the GCF of $20a$ and 8. Since 4 is the GCF of the numerator and the denominator, divide each one by that number. The result is $\frac{5a}{2}$.

6. B

The first step is to substitute the values of the variables you are given. Since $s = 4$, $2s = 8$. Since $r = 6$, $3r = 18$.

$$\frac{2s}{3r} = \frac{8}{18}$$

We can now simplify by dividing the numerator and the denominator by their GCF, which is 2.

$$\frac{8}{18} = \frac{8 \div 2}{18 \div 2} = \frac{4}{9}$$

(D) results from substituting the same value for both variables, and (E) results from substituting the value of s for r, and vice versa.

7. B

You need to simplify each fraction until you come upon the one with a value of $\frac{25}{3}$. That is a matter of finding the GCF for each numerator–denominator pair, and then dividing each both numbers by that value. Let's look at the prime factors of the numerator and denominator in Choice (B):

> 300: 2, 2, 3, 5, 5
>
> 36: 2, 2, 3, 3

Since the common prime factors are 2, 2, and 3, the GCF of 300 and 36 is 12. Let's test that number:

> $300 \div 12 = 25$
>
> $36 \div 3 = 12$
>
> $\frac{300}{36} = \frac{25}{3}$

(C) would result from multiplying the numerator 25, but from multiplying the denominator 20. (D) results from multiplying the numerator by 20, and the denominator by 15. (E) results from multiplying the numerator by 30, and the denominator by 12.

8.   D

Since the LCM of 6 and 8 is 24, convert both fractions so that they have denominators of 24.

$$\frac{2x}{6} = \frac{2x \cdot 4}{6 \cdot 4} = \frac{8x}{24}$$

$$\frac{3}{8} = \frac{3 \cdot 3}{8 \cdot 3} = \frac{9}{24}$$

$$\frac{8x}{24} + \frac{9}{24} = \frac{8x + 9}{24}$$

(A) results from multiplying the numerator of $\frac{2x}{6}$ by 3 instead of 4. (E) results from multiplying the numerator of $\frac{3}{8}$ by 4 instead of 3. (C) results from not multiplying the numerator of $\frac{3}{8}$ by 3.

CHAPTER 4

Roots, Radicals, and Powers

WHAT ARE ROOTS, RADICALS AND POWERS?

Roots, radicals, and powers are all related concepts in Algebra I. Many algebraic expressions feature variables multiplied by themselves.

Let's take the expression $2 \bullet 2 \bullet 2 \bullet 2$. You are likely to encounter expressions like this in Algebra I, but in a shorter form. Since 2 appears as a factor 4 times, we can write the expression as "2 to the 4th power", or 2^4. When an expression consists of the same factor being multiplied, for instance as $3 \bullet 3$ or $4 \bullet 4 \bullet 4 \bullet 4$, we call that expression a *power*. In the power 2^4, we call the number 2 the *base*, and the number 4 the *exponent*. You can always recognize an exponent by its size and position.

In Algebra I, the numbers 2 and 3 are used as exponents quite a lot; so often, in fact, that we have special terms for them. In a power with an exponent of 2, such as 3^2, we often say that the based is "squared." We may refer to 3^2 as "Three squared." In a power with an exponent of 3, we often say that the base is "cubed." So the expression 4^3 can be read as "four cubed." We use the term "squared" because the area of a square is the product of two equal dimensions (length and width), and we use the term "cubed" because the volume of a cube is the product of three equal dimensions (length, width, and height).

That brings us to roots. Since the expression 3^2, which is $3 \bullet 3$, has a value of 9, we could say that 3 squared is 9. The other side of the coin is that 3 is a *square root* of 9. 4 is a square root of the product of 4 and 4, 5 is a square root of $5 \bullet 5$, and so on. If you see the pattern here, you'll understand why we say that 6 is a square root of 36, and 7 is a square root of 49. Likewise, 2 is the *cube root* of 8, since 8 is the value of $2 \bullet 2 \bullet 2$.

Did you notice we said that 3 is *a* square root of 9? There is in fact another square root of 9: –3. After all, $(-3)^2$ is $-3 \bullet -3$, which also has a value of 9. Sometimes, however, we are only really interested in the positive square root of a number. In that case, we use a special sign, called a *radical sign*.

$$\sqrt{}$$

A *radical* is an expression with a number inside the sign:

$$\sqrt{4}$$

This expression, which you could call "radical 4" or the square root of 4, is the positive square root of 4. The radical sign looks like a division sign, but don't be fooled. The little "tail" on the left side of the sign makes it a different creature altogether!

CONCEPTS TO HELP YOU

Powers, roots, and radicals are closely related concepts, but each is important in its own right. You're likely to encounter questions in Algebra I about each of them. We're going to start by talking about powers, though, because it's easier to use those to make sense of roots. Roots are in turn important for understanding radicals. However, you'll also find that factoring is an important part of working with radicals.

Evaluating and Simplifying Expressions with Powers

Integers

A power is the product of a number multiplied by itself, however many times. A number raised to the 4th power is a product, with that number appearing as a factor four times; so 3^4 equals $3 \bullet 3 \bullet 3 \bullet 3$, or 81.

4^5 is the product of five factors of 4:

$$4 \bullet 4 \bullet 4 \bullet 4 \bullet 4 =$$
$$16 \bullet 4 \bullet 4 \bullet 4 =$$
$$64 \bullet 4 \bullet 4 =$$
$$256 \bullet 4 =$$
$$1,024$$

Negative numbers can be raised to powers as well. If the exponent is even, then the value of the power is even. If the exponent is odd, the power is odd.

$(-3)^2 = -3 \bullet -3 = 9$

$(-3)^3 = -3 \bullet -3 \bullet -3 = 9 \bullet -3 = -27$

$(-3)^4 = -3 \bullet -3 \bullet -3 \bullet -3 = 9 \bullet -3 \bullet -3 = -27 \bullet -3 = 81$

$(-3)^5 = -3 \bullet -3 \bullet -3 \bullet -3 \bullet -3 = 9 \bullet -3 \bullet -3 \bullet -3 = -27 \bullet -3 \bullet -3 = 81 \bullet -3 = -243$

NUMBERS RAISED TO THE FIRST POWER

Any number raised to the first power is equal to itself. So $2^1 = 2$ and $5^1 = 5$.

Variables

We've already seen examples where variables have been multiplied together. The product of x and y, for instance, is xy. But what if we want to multiply x by x? Do we write the product as xx? That is never done in Algebra. Instead, we use exponents to express the product. Just as $2 \bullet 2$ is 2^2, $x \bullet x$ is x^2. In fact, powers with variables play a big role in Algebra I, as you'll see in chapters 10 and 11.

Combining Powers

You can always multiply or divide powers, as long as the bases are the same: 2^2 by 2^3. Multiplying the powers is a matter of *adding the exponents*. Many people multiply the *exponents* instead. This is a very common mistake. Let's evaluate the following multiplication expression.

$2^2 \bullet 2^3 = 2^{2+3} = 2^5 = 32$

So why do we add the exponents? It would help to rewrite each power as a multiplication expression:

$2^2 = 2 \bullet 2$

$2^3 = 2 \bullet 2 \bullet 2$

$2^2 \bullet 2^3 = (2 \bullet 2) \bullet (2 \bullet 2 \bullet 2)$

Since the last expression includes 2 as factor fives times, it is equal to 2^5. To check, let's evaluate the powers before multiplying:

$$2^2 = 4$$
$$2^3 = 8$$
$$4 \bullet 8 = 32$$

And 32 is the value of 2^5!

To divide powers, subtract the exponents instead. Take 2^5 and 2^3. 32 divided by 8 is 4, and that is the value of 2^2. You could have gotten the exponent 2 by subtracting 3 from 5.

Remember that the bases must be the same. Though you can add exponents to find the product of 3^3 and 3^4, you cannot do that to find the product of 3^3 and 4^4.

All of this helps to explain another important concept: 0 as an exponent. All you need to know is that any number raised to the 0 power has a value of 1. So $2^0 = 1$, $3^0 = 1$, and so on.

Think about this way: since 4 divided by 4 has a value of 1, so does divided 2^2 by 2^2. But $2^2 \div 2^2 = 2^{2-2} = 2^0$. So $2^0 = 1$.

WRITING WITH AN EXPONENT OF 1

Any number can be rewritten with an exponent of 1; $2^1 = 2$, $3^1 = 3$, and so on. This comes in handy if you are asked to divide, say, 2^4 by 2. You can add the exponent of 1 to 2 to make it a power, and then subtract the exponents.

We now have a way of multiplying and dividing powers without evaluating them first. Unfortunately, there is nothing like this we can do to add and subtract powers.

Square Roots and Cube Roots

One key difference between square roots and cube roots is that only positive numbers can have square roots, but both positive and negative numbers can have cube roots. After all, there is no number— positive or negative—that you can square to get a negative number. When you cube a number, on the other hand, the result is negative if the number is negative, and it is positive if the number is positive:

$$(-2)^3 = -2 \bullet -2 \bullet -2 = 4 \bullet -2 = -8$$
$$3^3 = 3 \bullet 3 \bullet 3 = 9 \bullet 3 = 27$$

Try other examples for yourself. If the value of the power is negative, then the base is negative. This means that the cube root of a negative number is negative. Positive numbers, on the other hand, have both positive and negative *square* roots. Since $(-4)^2 = -4 \bullet -4 = 16$ and $4^2 = 4 \bullet 4 = 16$, both −4 and 4 are square roots of 16.

The square roots and cube roots we have been using as examples have been whole numbers. Those are easy examples to work with, but most square roots and cubes roots are not integers. Not only are the square roots of 2 and 3 not integers, you'd need to use a calculator to get their values to more than 1 or 2 decimal places.

The numbers that have integer square roots are called *perfect squares*. 25 is a perfect square because it has a square root of 5. 15, on the other hand, is not a perfect square, because its value is approximately 3.87.

Perfect squares are very important when it comes to the concept of radicals.

Simplifying and Evaluating Radicals

We often use radicals to express square roots that don't have integer values. A radical, you might recall, is an expression whose value is the positive square root of a number. Here are some examples:

$$\sqrt{9} = 3$$
$$\sqrt{16} = 4$$
$$\sqrt{100} = 10$$

The values of radicals of numbers that are not perfect squares can be given only approximately as decimals:

$$\sqrt{5} = 2.236...$$
$$\sqrt{8} = 2.828...$$

(By the way, these last decimals actually go on forever, without any pattern of repeating digits.)

Though we can't always evaluate radicals without using these decimals, we can often *simplify* them. Simplifying a radical is a matter of factoring, so that you wind up with an expression with the smallest possible number inside the radical sign. This involves finding the perfect square factors of a number.

Let's simplify $\sqrt{50}$. The key is to find a factor of 50 that's a perfect square. 1 is a factor of 50, but that's not helpful; we need numbers greater than 1. The next few perfect squares are 4, 9, 16, 25. Only that last number is a factor of 50. So let's rewrite our radical in terms of factors of 50:

$$\sqrt{50} = \sqrt{25 \cdot 2}$$

Now, when you have a radical of the product of two factors, it is equal to the product of the radicals of those two numbers:

$$\sqrt{A \cdot B} = \sqrt{A} \cdot \sqrt{B}$$

So we can say that:

$$\sqrt{25 \cdot 2} = \sqrt{25} \cdot \sqrt{2}$$

Since 5 is the positive square root of 25:

$$\sqrt{25} \cdot \sqrt{2} =$$
$$5 \cdot \sqrt{2} \text{ or } 5\sqrt{2}$$

As the other factor of 50, the 2 stays inside the radical sign. At this point $\sqrt{50}$ is fully simplified, since 2 has no perfect square factors (other than 1).

Here's another quick example:

$$\sqrt{12} = \sqrt{4 \cdot 3} = \sqrt{4} \cdot \sqrt{3} = 2\sqrt{3}$$

So the square root of 12 is the same value as twice the square root of 3.

It is also helpful to know that algebraic powers can be perfect squares.
$\sqrt{x^2} = x$, and $\sqrt{9y^2} = \sqrt{9} \bullet \sqrt{y^2} = 3 \bullet \sqrt{y^2} = 3y$

Multiplying and Dividing Radicals

Getting the product of two radicals is very straightforward. Take the product of the numbers inside the signs, and put it inside a radical.
$$\sqrt{5} \bullet \sqrt{6} = \sqrt{30}$$

If there are numbers outside the radical signs, get their product separately, and put it on the outside of the radical sign in the product.
$$2\sqrt{7} \bullet 3\sqrt{8} = 6\sqrt{56}$$

If only one of the radicals has a number on the outside, treat the other radical as if it had a 1 on the outside.
$$3\sqrt{5} \bullet \sqrt{2} = 3\sqrt{5} \bullet 1\sqrt{2} = 3\sqrt{10}$$

That's all there is to it. The division works in the same way.

Adding and Subtracting Radicals

To see how addition and subtraction work, lets work with a couple of numbers that can be rewritten as radicals. 2 can be rewritten as $\sqrt{4}$, and since $10 = 5 \bullet 2$, $10 = 5\sqrt{4}$. So what is the sum of $\sqrt{4}$ and $5\sqrt{4}$? Since those radicals equal 2 and 10, respectively, their sum is 12. 12 is the product of 6 and 2, and so $12 = 6\sqrt{4}$. We can see, then, that adding radicals need not involve changing the number inside the radical sign.

Adding radicals is similar to combining like terms in simplifying algebraic expressions. You can add radicals as long as the expressions inside the radical signs are the same. So you can simplify this expression:
$$4\sqrt{2} + \sqrt{2}$$

You can't simplify this expression, however, because the numbers inside the radicals are different:
$$2\sqrt{3} + \sqrt{2}$$

As long as the terms within the radical sign are the same, the radicals can be added: Just add the numbers on the outside and attach the sum to the radical:

$$4\sqrt{3} + 2\sqrt{3} =$$
$$(4 + 2)\sqrt{3} =$$
$$6\sqrt{3}$$

$4\sqrt{3}$, after all, is $(\sqrt{3} + \sqrt{3} + \sqrt{3} + \sqrt{3})$, and $2\sqrt{3}$ is $(\sqrt{3} + \sqrt{3})$, and $(\sqrt{3} + \sqrt{3} + \sqrt{3} + \sqrt{3}) + (\sqrt{3} + \sqrt{3}) = 6\sqrt{3}$

That should help to explain the result we got in the earlier example:

$$5\sqrt{4} + \sqrt{4} =$$
$$(5 + 1)\sqrt{4} =$$
$$6\sqrt{4}$$

Note that we used the distributive property to get $(4 + 2)\sqrt{3}$ from $4\sqrt{3} + 2\sqrt{3}$.

The same principle holds for subtraction:

$$5\sqrt{6} - 2\sqrt{6} =$$
$$(5 - 2)\sqrt{6} =$$
$$3\sqrt{6}$$

IF ONE RADICAL HAS A NUMBER OUTSIDE

If only one of the radicals has a number on the outside, treat the other radical as if it had a 1 on the outside:

$$6\sqrt{3} + \sqrt{3} = 6\sqrt{3} + 1\sqrt{3} =$$
$$7\sqrt{3}$$

Squaring Radicals

Since a radical is a positive square root of a number, squaring the radical gets you that number. So

$$\left(\sqrt{3}\right)^2 = 3$$
$$\left(\sqrt{5}\right)^2 = 5$$

In other words, a radical can be squared simply by getting rid of the radical sign.

STEPS YOU NEED TO REMEMBER

Think of powers and radicals as operations, to be handled along with addition, multiplication, etc., though with a bit more flexibility. Whether it's best to simplify early on depends on the situation.

Order of Operations

Since the order of operations requires you to evaluate powers first, you must do that before carrying out multiplication, addition, etc. Of course, you might have to do your substitutions first, if you're evaluating an algebraic expression.

Evaluating and Simplifying Terms

When it comes to adding and subtracting unlike radicals, you'll need to simplify at least one of them first. When it comes to multiplication and division, however, you should carry out the operation first, and then simplify.

Sometime a radical contains a term with more than one perfect square factor; you are not necessarily finished simplifying a radical after factoring out just one perfect square. Make sure that the term remaining inside the radical sign has no perfect square factors.

Using the Distributive Property to Combine

If you're ever in doubt as to how to add or subtract radicals, just remember the distributive property:

$$4\sqrt{3} + 2\sqrt{3} = (4 + 2)\sqrt{3}$$
$$= 6\sqrt{3}$$

Combine radicals is just like combining *like algebraic terms*.

STEP–BY–STEP ILLUSTRATION OF THE 5 MOST COMMON QUESTION TYPES

It's time to walk through some more common question types. Many of these questions also tie in concepts and topics from the earlier chapters. In Algebra I, you'll need to handle radicals and powers in algebraic expressions.

Question 1: Multiplying and Dividing Powers

What is the product of x^4 and x^3?

(A) x

(B) x^7

(C) x^{12}

(D) x^{64}

(E) x^{81}

Multiplying powers is just a matter of adding the exponents. Since $4 + 3 = 7$, $x^4 \bullet x^3 = x^7$. The best way to understand why you add the exponents instead of multiplying is to write out the powers as products:

$$x^4 = x \bullet x \bullet x \bullet x$$
$$x^3 = x \bullet x \bullet x$$
$$x^4 \bullet x^3 = (x \bullet x \bullet x \bullet x) \bullet x \bullet x \bullet x$$

Since x appears seven times in this multiplication expression, it is x^7. **So (B) is correct**.

When you divide the exponents, you subtract the exponents. (A) would be the result of dividing x^4 by x^3, since $4 - 3 = 1$. (C) results from multiplying the exponents, but this isn't what you do to get the product of two powers. In (D), the exponent was obtained by raising 4 to the 3rd power, and (E), it was obtained by raising 3 to the 4th power.

Question 2: Taking Roots

What is the cube root of –729?

(A) –243

(B) –27

(C) –9

(D) 9

(E)   27

The cube root of –729 is the number that has a value of –729 when raised to the third power. One thing we know is that the cube root of a negative number is always negative. So we can quickly rule out (D) and (E). Looking at (A), you might expect the value of –243 cubed to be much, much larger (in absolute value) than –729. In fact, its value is less than negative one million! The more practice you have, the more you'll see that certain numbers just can't possibly be cube or square roots of certain other numbers.

That leaves choices (B) and (C):

$$(-27)^3 = -27 \bullet -27 \bullet -27 = 729 \bullet -27$$

We can already see that $(-27)^2 = 729$; $(-27)^3$ has to have a greater absolute value, and so (B) is out.

$$(-9)^3 = -9 \bullet -9 \bullet -9 = 81 \bullet -9 = -729$$

So (C) is correct.

Question 3: Simplifying Radicals

Which of these is a *fully* simplified form of $\sqrt{108}$?

(A) $2\sqrt{27}$

(B) $3\sqrt{12}$

(C) $6\sqrt{3}$

(D) 12

(E)   18

You can always start by going through the perfect squares that are less than the number in the radical sign to find perfect square factors. You should find that 4 is a factor of 108, as 108 is the product of 4 and 27. So:

$$\sqrt{108} =$$
$$\sqrt{4 \cdot 27} =$$
$$\sqrt{4} \cdot \sqrt{27} =$$
$$2\sqrt{27}$$

Though this is choice (A), that is not the correct answer, since the radical can be simplified further. That's because 9, a perfect square, is a factor of 27. So:

$$2\sqrt{27} =$$
$$2\sqrt{9 \cdot 3} =$$
$$2 \cdot \sqrt{9} \cdot \sqrt{3} =$$
$$2 \cdot 3 \cdot \sqrt{3} =$$
$$6\sqrt{3}$$

So **(C) is the correct answer.** (B) is also on the right track, but $3\sqrt{12}$ is not fully simplified; you can still factor 4 out of 12 and simplify $\sqrt{4}$. Choices (D) and (E), on the other hand do not have the same value as $\sqrt{108}$. (D) might be the result of incorrectly taking 12 to be the square root of 108, while (E) might result from a mistaken attempt to evaluate the product of 6 and $\sqrt{3}$.

Question 4: Evaluating Radicals

If $a = 3$ and $b = 4$, what is the value of $\sqrt{a^4 b^3}$?

(A) 12

(B) 24

(C) 36

(D) 72

(E) 144

Your first step here is to substitute the values of the variables you are given.

$$\sqrt{a^4 b^3} = \sqrt{3^4 \bullet 4^3}$$

The next step in the order of operations is to evaluate powers. Since $3^4 = 81$ and $4^3 = 64$:

$$\sqrt{3^4 \bullet 4^3} = \sqrt{81 \bullet 64}$$

Now you could perform the operation inside the radical sign, but that would be going in the wrong direction. Instead, break up the radical into parts we can then evaluate.

$$\sqrt{81 \bullet 64} =$$
$$\sqrt{81} \bullet \sqrt{64} =$$
$$9 \bullet 8 =$$
$$72$$

(D) is the correct answer. The other choices all result from evaluating the wrong powers. 12 is the value of $\sqrt{a^2 b^2}$, 24 is the value of $\sqrt{a^3 b^2}$, 36 is the value of $\sqrt{a^4 b^2}$, and 144 is the value of $\sqrt{a^4 b^4}$. Each of these errors could come from raising 3 or 4 to the wrong power.

Question 5: Combining Radicals

What is $5\sqrt{3} + \sqrt{2} \bullet \sqrt{96} =$

(A) $13\sqrt{3}$

(B) $14\sqrt{3}$

(C) $5\sqrt{3} + 14$

(D) 39

(E) 117

Here, the multiplication must be carried out first. Since the product of two radicals is the radical of the product of the numbers inside the signs, $\sqrt{2} \bullet \sqrt{96} = \sqrt{192}$. Now we cannot simplify $5\sqrt{3} + \sqrt{192}$ without first simplifying $\sqrt{192}$. Since 192 is the product of 4 and 48:

$$\sqrt{192} =$$
$$\sqrt{4 \cdot 48} =$$
$$\sqrt{4} \cdot \sqrt{48} =$$
$$2\sqrt{48}$$

Now we need to try to simplify $2\sqrt{48}$ further. Since 48 is the product of 16 and 3:

$$2\sqrt{48} =$$
$$2\sqrt{16 \cdot 3} =$$
$$2\sqrt{16} \cdot \sqrt{3} =$$
$$2 \cdot 4 \cdot \sqrt{3} =$$
$$8\sqrt{3}$$

You might have also found that 192 is the product of 64 and 3, and then simplified $\sqrt{64 \cdot 3}$ to get the same result.

$5\sqrt{3} + \sqrt{192}$, then, is equal to $5\sqrt{3} + 8\sqrt{3}$.

Remember that you can add two radical expressions by adding the numbers in front (as long as the expressions have *like* radicals): $5\sqrt{3} + 8\sqrt{3} = 13\sqrt{3}$.

So (A) is correct. (C) results from getting 196 instead of 192 as the product of 2 and 196. (D) results from taking the sum of $5\sqrt{3}$ and $8\sqrt{3}$ to be the sum of 13 and the product of $\sqrt{3}$ and $\sqrt{3}$.

CHAPTER QUIZ

1. Which power has a value of 64?
   - (A) 2^5
   - (B) 2^6
   - (C) 4^4
   - (D) 8^3
   - (E) 4^5

2. $y^8 \div y^2 =$
   - (A) y^3
   - (B) y^4
   - (C) y^6
   - (D) y^{10}
   - (E) y^{16}

3. Which of the following numbers is a perfect square?
   - (A) 2
   - (B) 3
   - (C) 5
   - (D) 8
   - (E) 9

4. $\sqrt{32} - \sqrt{2} =$
   - (A) $3\sqrt{2}$
   - (B) $\sqrt{30}$
   - (C) $5\sqrt{2}$
   - (D) $10\sqrt{3}$
   - (E) $15\sqrt{2}$

5. $\sqrt{8} \bullet \sqrt{18} =$
   - (A) $\sqrt{26}$
   - (B) $6\sqrt{2}$
   - (C) $4\sqrt{6}$
   - (D) 12
   - (E) 14

6. $\sqrt{147x^3} =$
   - (A) $7x\sqrt{3x}$
   - (B) $7x^2\sqrt{3x^2}$
   - (C) $7x^2\sqrt{3}$
   - (D) $7x^2\sqrt{3x}$
   - (E) $7x\sqrt{3x^2}$

7. If $x = 4\sqrt{5}$, $x\sqrt{20} =$
   - (A) 10
   - (C) 20
   - (B) 25
   - (D) 40
   - (E) 100

8. If s = 3 and t = 5, which radical equals $3\sqrt{3}$?
   - (A) $\sqrt{3s + 2t}$
   - (B) $\sqrt{5s + t}$
   - (C) $\sqrt{2s + 4t}$
   - (D) $\sqrt{4s + 3t}$
   - (E) $\sqrt{s + 5t}$

ANSWER EXPLANATIONS

1. B

Here we just need to evaluate the powers to see which has the given value:

$2^5 = 2 \bullet 2 \bullet 2 \bullet 2 \bullet 2 = 4 \bullet 2 \bullet 2 \bullet 2 = 8 \bullet 2 \bullet 2 = 16 \bullet 2 = 32$

$2^6 = 32 \bullet 2 = 64$

$4^4 = 4 \bullet 4 \bullet 4 \bullet 4 = 16 \bullet 4 \bullet 4 = 64 \bullet 4 = 256$

$8^3 = 8 \bullet 8 \bullet 8 = 64 \bullet 8 = 512$

$4^5 = 256 \bullet 4 = 1{,}024$

Since $2^6 = 64$, Choice (B) is the answer. Notice the simplified way in which we evaluated 2^6. Since $2^5 = 32$ and $2^6 = 2^5 \bullet 2$, we just multiplied 32 by 2.

2. C

You can divide powers by *subtracting* exponents, as long as the bases of the powers are the same. Since $8 - 2 = 6$, $y^8 \div y^2 = y^6$, and (C) is the answer. (B) is tempting, since 4 is the result of dividing 8 by 2. But since you're not supposed to divide the exponents, it is not correct. (D) is the product of the two powers, and (E) might be an attempt to get the product by multiplying the exponents.

3. E

To check whether a number is a perfect square, square some integers that are lower than it. Take the number 3: we find that the number 1 squared is 1, and 2 squared is 4. Since there are no integers between 1 and 2, you can be sure that no integer squared equals 3.

Of all the numbers on the list, only 9 has integer square roots (3 and –3), so (E) is the answer. (D) is in fact a perfect cube, but it is not a perfect square. (A) might be tempting for someone who associates the number 2 with "squared," but that number is not a perfect square.

4. A

To combine the radical expressions, you'll need to simplify $\sqrt{32}$:

$\sqrt{32} = \sqrt{16 \bullet 2} = \sqrt{16} \bullet \sqrt{2} = 4\sqrt{2}$

So $\sqrt{32} - \sqrt{2} = 4\sqrt{2} - \sqrt{2}$

We can also write out $4\sqrt{2} - \sqrt{2}$ as $4\sqrt{2} - 1\sqrt{2}$

Subtracting those radical is then a matter of subtracting the numbers outside the radical signs, since terms have like radicals:

$4\sqrt{2} - \sqrt{2} = 3\sqrt{2}$

(B) results from subtracting the numbers inside the radical signs. (C) is what you might get if you took 6 instead of 4 to be a square root of 16. (E) is the result of getting $16\sqrt{2}$ instead of $4\sqrt{2}$ from $\sqrt{32}$; remember that you have to use the square root of 16.

5. D

Multiplying radicals like these is just a matter of multiplying the numbers inside the radical signs, and then putting the product inside a radical sign. Since the product of 8 and 18 is 144, $\sqrt{8} \bullet \sqrt{18} = \sqrt{144}$,

and since 12 is the positive square root of 144, $\sqrt{144} = 12$

(A) is the result of confusion of multiplication of radicals with multiplication of powers, as the multiplication of powers involves addition. (B) is the result of trying to multiply $2\sqrt{2}$ by $3\sqrt{2}$ without using the product of the radicals.

6. A

You can begin here by factoring perfect squares out of $147x^3$. Since 147 is the product of 49 and 3:

$\sqrt{147x^3} = \sqrt{49 \bullet 3x^3}$

Since $\sqrt{49 \bullet 3x^3} = \sqrt{49} \bullet \sqrt{3x^3}$, we can simplify our radical to get $7\sqrt{3x^3}$. Since $x^2 \bullet x = x^3$, x^2 is a factor of $3x^3$:

$7\sqrt{3x^3} = 7\sqrt{3 \bullet x^2 \bullet x} = 7 \bullet \sqrt{x^2} \bullet \sqrt{3x}$

Since $\sqrt{x^2} = x$:

$7 \bullet \sqrt{x^2} \bullet \sqrt{3x} = 7x\sqrt{3x}$

(C) results from taking x^2 to be the square root of x^3. (D) involves factoring out x^2, without taking its square root.

7. D

First we need to replace the variable x with the value given.

$x\sqrt{20} =$

$4\sqrt{5} \bullet \sqrt{20}$

Since $5 \bullet 20 = 100$, the product of $\sqrt{5}$ and $\sqrt{20}$ is $\sqrt{100}$, and

$4\sqrt{5} \bullet \sqrt{20} = 4\sqrt{100}$

$\sqrt{100}$ is the positive square root of 100, which is 10, so

$4\sqrt{100} =$

$4 \bullet 10 =$

40

You might have gone with (A) if you overlooked the 4 on the outside of the radical. (B) would be the result of getting $\sqrt{25}$ instead of $\sqrt{100}$ when multiplying $\sqrt{5}$ and $\sqrt{20}$. (C) would be tempting if you take $4\sqrt{5}$ to equal $\sqrt{20}$.

8. D

Test the options by plugging in values and simplifying until you find one that equals $3\sqrt{3}$. It might be easier to find the right expression in this case by working backward. Since $3 = \sqrt{9}$, $3\sqrt{3} = \sqrt{9} \bullet \sqrt{3} = \sqrt{27}$.

So we are look for an expression inside a radical that has a value of 27. Let's check them out:

$3s + 2t = 3 \bullet 3 + 2 \bullet 5 = 9 + 10 = 19$

$5s + t = 5 \bullet 3 + 5 = 15 + 5 = 20$

$2s + 4t = 2 \bullet 3 + 4 \bullet 5 = 6 + 20 = 26$

$4s + 3t = 4 \bullet 3 + 3 \bullet 5 = 12 + 15 = 27$

$s + 5t = 3 + 5 \bullet 5 = 3 + 25 = 28$

Since $4s + 3t = 27$, (D) is correct.

CHAPTER 5

Equation Solving

WHAT IS EQUATION SOLVING?

An *equation* is a mathematical statement that relates two expressions as equal. Equations such as these do not involve algebraic expressions:

$$4 + 5 = 9$$
$$\frac{3}{5} = \frac{9}{15}$$
$$\sqrt{9} = 3$$

You have already evaluated algebraic expressions; basic equation solving is like working backward. In the most basic kind of equation solving, you start with the value of an algebraic expression, which you use to find the value of the variable.

Take the equation $2x = 10$. You need to figure out the value of x, given what you know. That x multiplied by 2 equals 10. The only number that gets you 10 when multiplied by 2 is 5. So x must have a value of 5.

You can also say that 5 is the *solution of the equation*. The solution of the equation is that value of the variable that makes it *true*.

So the solution of $3x = 12$ is 4; that's because the equation is true when $x = 4$.

CONCEPTS TO HELP YOU

Algebraic expressions such as $3x = 12$ are as simple as they come. But many other kinds of equations require several steps to solve. The concepts we will review here are important for mastering those steps.

The key to basic equation solving is to *get the variable alone on one side* of the equation. If you manipulate an equation correctly, you can rewrite it such a way that the variable appears by itself on just one side of the equation (the two sides are separated by the '=' sign). Once you have achieved that, the solution should be within your grasp.

Addition, Subtraction, Multiplication, and Division Properties of Equality

This is an important concept. Let's start with a simple equation: $2^2 = 4$. Suppose you add 3 to each of the expressions: $2^2 + 3 = 4 + 3$. Now both sides equal 7. In fact, whenever you add the same number to both sides of an equation that is true, you get a new equation that is true.

Since $\sqrt{9} = 3$, for example, we also know that $\sqrt{9} + 1 = 3 + 1$. Check for yourself that each side of the equation has a value of 4.

This also holds for algebraic equations. Suppose $x + 2 = 7$. If so, then $(x + 2) + 5 = 7 + 5$. (We added 5 to each side.) Simplifying each side gets us $x + 7 = 12$, so you can rest assured that if $x + 2 = 7$, then $x + 7 = 12$.

As long as you correctly add the same number to both sides, you can't go wrong. This is known as the *Addition Property of Equality*. It has its counterparts in subtraction, multiplication, and division. Some examples:

Subtraction Property of Equality
If $3x + 5 = 8$,
$(3x + 5) - 2 = 8 - 2$;
so $3x + 3 = 6$.

Multiplication Property of Equality
If $x + 1 = 4$,
$(x + 1) \bullet 3 = 4 \bullet 3$;
so $3x + 3 = 12$.

Division Property of Equality
If $4x = 8$,
then $4x \div 2 = 10 \div 2$;
so $2x = 5$.

Getting the Variable Alone

There actually isn't much of a point to performing the operations seen above; they are just examples. Instead, we'll want to perform operations that cause numbers to be *eliminated* from one side of the equation.

Suppose $x + 4 = 10$. You might be able to solve this equation in your head, but use the *Subtraction Property of Equality* instead. Let's subtract 4 from each side of the equation:

$$x + 4 = 10$$
$$x + 4 - 4 = 10 - 4$$
$$x = 6$$

Now the solution of $x + 4 = 10$ is leaping off the page. Once you have gotten the variable alone on one side, you have solved the original equation. The solution is 6.

The important thing is knowing what to do to both sides of an equation so you can isolate the variable on one side. When a number is being added to the variable on one side of the equation, you should subtract that same number. The two numbers "cancel each other out." Just remember that when you perform an operation on one side of an equation, *you must perform the very same operation on the other side*.

This is where the concepts of the additive and multiplicative inverses come in handy. Subtracting the number that is being added is like combining it with its additive inverse. That's why the two numbers cancel out—since the sum of a number and its additive inverse is 0.

Let's look at an example involving multiplication and division. Suppose $5x = 35$. We would like to perform an operation that would get rid of the number 5 on the left side. What should we do? Here's a hint: $x = 1 \bullet x$.

We might carry out an operation on $5x$ that gets us $1x$. Since $5x \div 5 = 1x$, we should divide $5x$ by 5, and do the same to the right side of the equation as well: $5x \div 5 = 35 \div 5$. The result is $x = 7$, and so 7 is the solution of $5x = 35$. You can check that by finding that the product of 5 and 7 is 35.

So when an equation involves a variable being multiplied by a number, division by the same number "cancels it out." Think back to the concept of the multiplicative inverse covered in chapter 1. Multiplying a number by its multiplicative inverse gives you a product of 1, which is the result you'd get by dividing the number by itself. So when you have a variable multiplied by a number, you can get rid of the number by multiplying the expression by the number's multiplicative inverse; that is the same thing as dividing. Consider this equation:

$$\frac{x}{2} = 6$$

The variable x is being divided by 2, or multiplied by $\frac{1}{2}$. Those amount to the same thing! Let's say that it is divided by 2. The solution lies, then, in canceling out the division with multiplication. Let's multiply both sides by 2:

$$\left(\frac{x}{2}\right) \bullet 2 = 6 \bullet 2$$

$$\frac{2x}{2} = 12$$

A fraction can be simplified by dividing top and bottom by its GCF. In that case,

$$\frac{2x}{2} = \frac{1x}{1} = x$$

So $x = 12$.

We can sum up it like this :

–Addition cancels out subtraction

–Subtraction cancels out addition

–Division cancels out multiplication

–Multiplication cancels out division

Combining Like Terms

In many equations, variables can be found on both sides.

$$4x - 3 = 3x + 5$$

Getting the variable x alone here involves getting it on *one side only*. So you must subtract one of the variable terms from both sides. You know that terms with variables can be combined with addition or subtraction as long as they are *like terms*, so let's subtract $3x$ from both sides:

$$4x - 3 - 3x = 3x + 5 - 3x$$

Since $4x - 3x = x$ and $3x - 3x = 0$, the result is $x - 3 = 5$.

STEPS YOU NEED TO REMEMBER

We've been looking at equations that can be solved in just one step. Now we move on to equations that require two or more steps to solve, which are far more common.

Order of Operations

Take the equation $3x + 6 = 18$. Getting the variable x alone will involve subtraction and division; we'll subtract 6 (because it is added on that side), and we'll divide by 3 (because the variable is multiplied by that number). But which step do we take first?

The general rule is that you follow the order of operations, but *in reverse*. In the original order of operations, you carry out multiplication and division before addition and subtraction. When it comes to equation solving, you do addition and subtraction first.

So let's first subtract 6 from both sides:

$$3x + 6 - 6 = 18 - 6$$

That simplifies to

$$3x = 12$$

Now we divide both sides by 3:

$$3x \div 3 = 12 \div 3$$

And that simplifies to

$$x = 4$$

So the solution to $3x + 6 = 18$ is 4.

Look for the "Hidden Parentheses"

At times it can appear that the order of operations isn't strictly followed. For example, the standard way to solve $\frac{x+2}{5} = 3$ is to first multiply both sides by 5, and then subtract 2 (from both sides). Does that go against the order of operations? The answer is no.

The original order of operations had you evaluate operations inside parentheses before all other multiplication and division. That's important, because the equation we are dealing with here involves "hidden" parentheses.

$\frac{x+2}{5}$ can be rewritten as $(x + 2) \div 5$. Once you see that, you should see that following the reverse order of operations requires you to multiply before you subtract.

Checking Your Work

Once you have a solution for an equation, you can plug the value back into the original equation to make sure it's correct. Evaluate both sides of the equation, and see whether you get the same number.

Take $4x - 7 = 29$. If you had solved this correctly, you got 9 for the solution. Now take $4x - 7$ and evaluate it with x having a value of 9. You should get a value of 29 for $4x - 7$. Since that is the number on the right side of the equation, the solution checks out.

STEP–BY–STEP ILLUSTRATION OF THE 5 MOST COMMON QUESTION TYPES

Now it's time to go through some questions of the kinds you're likely to encounter elsewhere. These run from the basic to the more advanced. We'll revisit some concepts from earlier chapters, such as absolute values and radicals.

Question 1: Basic Equation Solving—"Variables on One Side"

If $2x + 5 = 21$, what is the value of x?

(A) 8

(B) 9

(C) 11

(D) 13

(E) 14

In this question, we need to carry out two operations—subtraction and division—though it's considered a basic equation because there is only one term with a variable.

The key to solving an equation is getting the variable alone on one side; here, the x should be on one side with no other numbers at all. Each operation we perform is meant to "move numbers away." We'll whittle down that left side of the equation until nothing but the variable remains.

Now recall you must follow the order of operations *in reverse* when solving an equation in two or more steps. The first step, then, is subtraction.

$$2x + 5 = 21$$

$$\underline{-5 \quad -5} \qquad \text{[5 is subtracted from both sides of the}$$
$$\text{equation]}$$

$$2x = 16$$

$$\div 2 \quad \div 2 \qquad \text{[Both sides are divided by 2]}$$

$$x = 8$$

So the solution of $2x + 5 = 21$ is 8, and (A) is correct. Look at the first step involving subtraction. A common mistake involves carrying out addition instead, since that is what already appears. If you add 5 to 21 instead of subtracting, you get 26. Half of that is 13, which is choice (D). You might have gotten 14 as a solution if you subtracted 2 in the last step instead of dividing by that number. Division always "cancels out" multiplication.

Question 2: Advanced Equation Solving—"Variables on Both Sides"

If $\dfrac{y+3}{4} = y - 6$, $y =$

(A) 3

(B) 7

(C) 9

(D) 12

(E) 15

Here, you have variables on both sides of the equation, and the expression on the left has "hidden parentheses." When an expression like $y + 3$ is divided by a number, treat it as though it were inside parentheses. The first step should be multiplying both side of the equation by 4, since the expression with the variable is divided by that number. You'll perform the operation that gets rid of the 3 later.

Multiply both sides by 4:

$$4\left(\frac{y+3}{4}\right) = 4(y-6)$$

$y + 3 = 4y - 24$	Subtract y from both sides
$3 = 3y - 24$	Add 24 to both sides to eliminate the 24 from the right side
$27 = 3y$	Divide both sides by 3
$y = 9$	

The solution of $\dfrac{y+3}{4} = y - 6$ is 9, and (C) is the answer. Performing the steps in the correct order is critical. If you tried to first subtract 3 from both sides and got $\dfrac{y}{4} = y - 9$, you most likely went on to get a solution of 12 (choice (D)).

Question 3: Solving an Equation for One of Several Variables

What is the solution of $3y + 4 = 5x - 2$ for y?

(A) $y = \dfrac{5x}{3} - 6$

(B) $y = \dfrac{5x - 14}{3}$

(C) $y = \dfrac{5x}{3} - 2$

(D) $y = \dfrac{5x + 2}{3}$

(E) $y = \dfrac{5x}{3} + 2$

Many equations involve different variables, yet when one equation alone has two different variables, however, you won't be able to find the values of both. To solve for two variables, you need at least two equations. We'll work with this later in the book, but for now, we'll solve a two-variable equation "for one variable." This is the same thing as solving an equation "in terms of one variable." You will see those phrases a lot, so be prepared.

To solve an equation "for y" is to write an equation that gives the *value of y in terms of x*. Solving for y, then, is a matter of getting that variable alone on one side of the equation.

In this example, since 4 is added on the left side, begin by subtracting 4 from both sides.

$$3y = 5x - 6 \qquad \text{Divide both sides by 3}$$
$$y = \frac{5x}{3} - 2$$

This is the solution "for y," so **(C) is the answer.** Watch out for questions that include answer choices providing solutions *for the other variable*. Remember your terminology, and you'll avoid picking such incorrect choices. You might have gotten (A), $y = \frac{5x}{3} - 6$, if you forgot to distribute the 3 when dividing in the last step. Since dividing by a number is just like multiplying by its multiplicative inverse, the distributive property holds.

Question 4: Absolute Value Equations

What are the solutions of $|x + 6| = 10$?

(A) −16 and −4

(B) −16 and 4

(C) −4 and 4

(D) 16 and 4

(E) 16 and −4

Absolute value equations are special because they can have more than one solution. This is because two different numbers—or two different algebraic expressions—can have the same absolute value. −2 and 2 have the same absolute value, as do $x + 6$ and $-(x + 6)$.

Now suppose that $x + 6$ has a positive value. In that case, $|x + 6| = 10$ if $x + 6 = 10$. But if $x + 6$ has a negative value, then $|x + 6| = 10$ if $-(x + 6) = 10$.

Because each of those equations has a different solution, the variable x has two possible values. To find them, we must solve each equation. Let's look at $x + 6 = 10$ first. All we have to do is subtract 6 from both sides to get $x = 4$. So, one of the solutions is 4.

As for $-(x + 6) = 10$, you can begin by getting rid of the negative sign in front of $x + 6$. You could do this by distributing it inside the parentheses, as we explained in Chapter 2. However, It would be much easier to simply multiply both sides of the equation by -1, since the product of two negatives is a positive:

$$-1 \bullet -(x + 6) = -1 \bullet 10$$
$$x + 6 = -10 \qquad \text{Subtract 6 from both sides}$$
$$x = -16$$

So the solutions are 4 and -16. Choice (B) is the answer.

Finding the second solution is a matter of solving the equation with the negative of the expression within the absolute value sign. The solutions of -4 and 4 would be tempting if you were looking for the negative of the first solution. Similarly, 16 and 4 in (D) would be tempting if you found the second solution, but then took its absolute value.

Question 5: Solving Equations with Radicals and Exponents

If $\sqrt{x + 5} = 8$, $x =$

(A) 3

(B) 9

(C) 59

(D) 69

(E) 169

Though this equation does not include any exponents, you may use one in solving it. There is nothing you can do with this equation as long as the radical sign is present. So how do we get rid of the sign?

When you square a radical, you wind of with the expression inside the sign:

$$\left(\sqrt{x+5}\right)^2 = x + 5$$

We can square $\sqrt{x+5}$ by multiplying the left side of the equation by that expression, as long as we do the same on the other side. But multiplying both sides by $\sqrt{x+5}$ won't really help us:

$$\left(\sqrt{x+5}\right)\sqrt{x+5} = \left(\sqrt{x+5}\right)8$$
$$x + 5 = \left(\sqrt{x+5}\right)8$$

We're not any closer to a solution with this result. Fortunately, there's something else we can do. Rather than multiply both sides by $\sqrt{x+5}$, let's multiply one side by $\sqrt{x+5}$, and the other by 8, since the equation we started with says that the two expressions are equal.

$$\left(\sqrt{x+5}\right)\sqrt{x+5} = 8 \bullet 8 \qquad \text{Simplify both sides of the equation}$$
$$x + 5 = 64 \qquad \text{Subtract 5 from both sides}$$
$$x = 59$$

(C) is correct here. If you added 5 to the right side instead of subtracting, you would have gotten 69 (Choice (D). You might have gotten 9 as a solution if you tried to subtract 5 from both sides before squaring the radical. Subtracting 5 from the right side would leave you with a value of 3 that one might go on to square. You cannot take numbers out of a radical by addition or subtraction.

CHAPTER QUIZ

1. What is the solution of $x + 13 = 9$?

2. If $-2k = 14$, then $k =$
   - (A) -12
   - (B) -7
   - (C) 7
   - (D) 12
   - (E) 16

3. If $6a - 9 = 57$, what is the value of a?
   - (A) 8
   - (B) 9
   - (C) 10
   - (D) 11
   - (E) 12

4. If $5x - 8 = 8x + 19$, $x =$
   - (A) -13
   - (B) -11
   - (C) -9
   - (D) 9
   - (E) 13

5. If $3y + 10 = 12x + 2$, $x =$
   - (A) $\dfrac{y}{4} + \dfrac{2}{3}$
   - (B) $\dfrac{y + 3}{4}$
   - (C) $\dfrac{y}{4} + 1$
   - (D) $\dfrac{y + 2}{3}$
   - (E) $\dfrac{y}{3} + \dfrac{3}{4}$

6. What is the solution of $4x - 7y + 12 = 20x + y - 4$ for y?
   - (A) $y = 1 - 2x$
   - (B) $y = 2 - 2x$
   - (C) $y = 2x + 1$
   - (D) $y = 3x + 2$
   - (E) $y = 3x - 2$

7. Which pair of numbers are both solutions of $|2x - 8| = 34$?
   - (A) -21 and 21
   - (B) -13 and 21
   - (C) -13 and 13
   - (D) 13 and -21
   - (E) 13 and 21

8. If $x^3 = -\sqrt{729}$, $x =$

ANSWER EXPLANATIONS

1. −4

Since 13 is added to the variable, you need to subtract that value from both sides:

$$x + 13 = 9$$
$$\underline{-13 \quad -13}$$
$$x = -4$$

The solution, −4, is the result of subtracting 13 from 9 on the right side. Don't be tempted by 4, which you would get by subtracting 9 from 13. And don't be tempted by 22, the sum of 9 and 13.

2. B

Here, the variable expression involves multiplication instead of addition, so you must divide in order to get the variable alone.

$$-2k = 14$$
$$\div -2 \div -2$$
$$k = -7$$

The solution of −7 is what you get by dividing 14 by −2. 12 and 16 are tempting choices because they are, respectively, the results of adding and subtracting −2 instead of dividing. Remember that division is the operation that cancels out multiplication.

3. D

The (reverse) order of operations requires you to handle addition and subtraction first. So start by adding 9 to both sides:

$$6a - 9 = 57$$
$$+9 \quad +9$$
$$6a = 66$$

Now the variable is just multiplied by 6; divide both sides by that number:

$$6a = 66$$
$$\div 6 \quad \div 6$$
$$a = 11$$

Choice (A), 8, would result from subtracting 9 from 57 instead of adding to get 48 instead of 66.

4. C

It takes three steps to get to this equation's solution. The first two steps involve combining like terms, so that there is only one variable term in the equation, which is alone on one side.

$$5x - 8 = 8x + 19$$

$\underline{\quad + 8 \quad\quad + 8 \quad}$ Add the same number to both sides

$$5x = 8x + 27$$

$\underline{-8x \quad -8x \quad}$ Subtract the same expression from both sides

$$-3x = 27$$

Now that we have a single variable term and it is alone on one side, we can carry out the last step:

$$-3x = 27$$

$\div -3 \div -3$ Divide both sides by the same number to get the variable alone

$$x = -9$$

Remember that you must divide by a negative number what the variable is multiplied by one. If you divide by the additive inverse of –3, you would arrive at 9 instead of –9.

5.   A

This question involves an equation with two variables. You're asked to solve it for one variable, x. That is the same as solving it "in terms of y."

$$3y + 10 = 12x + 2$$
$$\underline{-2 \qquad -2}$$
$$3y + 8 = 12x$$

Since our goal is to get x alone, we need to cancel out the number in the operation involving x. So we must divide both sides by 12, because x is multiplied by that number:

$$3y + 8 = 12x$$
$$\div 12 \qquad \div 12$$
$$\frac{3y+8}{12} = x$$
$$\frac{3y+8}{12} = \frac{3y}{12} + \frac{8}{12} \, .$$

Since $\frac{3y}{12}$ can be simplified to $\frac{y}{4}$ and $\frac{8}{12}$ can be simplified to $\frac{2}{3}$, $\frac{y}{4} + \frac{2}{3}$ is the solution of the equation for x. You would have gotten $\frac{y}{4} + 1$ instead if you added 2 to the left side of the equation instead of subtracting. You would have gotten $\frac{y+3}{4}$, (B), if you took $\frac{8}{12}$ to equal $\frac{3}{4}$ instead of $\frac{2}{3}$.

6.   B

It will take four steps to solve this equation. The order in which you combine like terms doesn't matter here. Start with the x terms:

$$4x - 7y + 12 = 20x + y - 4$$
$$\underline{-4x -4x}$$
$$-7y + 12 = 16x + y - 4$$

Subtract y from both sides to get

$-8y + 12 = 16x - 4$

Since you are supposed to solve the equation for y, you need to get the numbers off of the left side of the equation. Start with 12:

$$\begin{aligned} -8y + 12 &= 16x - 4 \\ -12 \qquad &\quad -12 \\ \hline -8y &= 16x - 16 \end{aligned}$$

Dividing both sides of this equation by -8 gets you, $y = -2x + 2$. Since $-2x + 2 = 2 - 2x$, that is the solution for y. You might have gotten $y = 1 - 2x$ instead if you subtracted incorrectly in the next to last step, and got -8 instead of -16 for the value of $-4 - 12$. You might have gotten a solution involving $3x$ instead of $2x$ if you added $4x$ to the right side instead of subtracting.

7. B

Solving an absolute value equation is like solving two equations: one involving the expression inside the absolute value sign, and another involving its negative. So you have to solve both $2x - 8 = 34$ and $-(2x - 8) = 34$.

First, solve $2x - 8 = 34$:

$$\begin{aligned} 2x - 8 &= 34 \\ +8 \quad &\quad +8 \\ \hline 2x &= 42 \\ \div 2 \quad &\quad \div 2 \\ x &= 21 \end{aligned}$$

Now solve $-(2x - 8) = 34$. To get rid of the negative sign, you can start by multiplying both sides by -1. You could instead distribute the negative sign, but multiplying will save you some work in this case.

$$-(2x - 8) = 34$$
$$\underline{-1 \qquad -1}$$
$$2x - 8 = -34$$
$$\underline{+8 \ +8}$$
$$2x = -26$$
$$\div 2 \ \div 2$$
$$x = -13$$

The solutions to $2x - 8 = 34$ and $-(2x - 8) = 34$ are 21 and –13, respectively. This means that both numbers are solutions to $|2x - 8| = 34$. Choice A, –21 and 21, is tempting, since one number is the absolute value of the other. The same goes for –13 and 21 in (C). You might have picked (E) if you took the absolute value of each solution. That step is not part of the solution process, however.

8.   –3

Solving this equation involves working with powers and radicals. Your first step should be to evaluate the radical. Don't be thrown off by the negative sign on the right side. Though you cannot evaluate radicals of negative numbers, the negative sign is outside the radical; ignore it for the moment, and evaluate $\sqrt{729}$. Since the positive square root of 729 is 27, $\sqrt{729} = 27$, and $-\sqrt{729} = -27$.

$$\text{So } x^3 = -27.$$

Now, since x is the cube root of x^3, it is the cube root of –27. Since $(-3)^3$ $= -3 - 3 - 3 = -27$, –3 is the cube root of –27, and that is the solution of $x^3 = -\sqrt{729}$.

One common incorrect answer to this question is –9. That comes as a result of ignoring the role of the radical sign, and taking the negative of the cube root of 729.

CHAPTER 6

Coordinate Geometry

WHAT IS COORDINATE GEOMETRY?

It's time now to introduce the *coordinate plane* and explain how it is used to graph points, lines, and figures. A *plane* is like a flat surface. In this case, it is one that goes on in all directions forever. *Coordinates* are numbers used to pinpoint locations on the plane; they are used in coordinate geometry to identify points on the coordinate plane (also referred to as the *coordinate grid*.)

CONCEPTS TO HELP YOU

Understanding the coordinate system is the key to working with key algebra concepts in coordinate geometry. These include equations that allow us the find the slope, midpoints, and lengths of line segments.

The Coordinate Plane

The number line is a line extending in two directions, containing all positive and negative numbers, with 0 in the center. Points can be plotted on this line. In the example below, a point is plotted at 3:

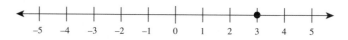

Here, we can plot points representing numbers along a single dimension. But suppose we wanted to use numbers to plot points in two dimensions? We might want to pinpoint locations on a map or a computer screen, perhaps. Picture, then, a second number line that is vertical and intersects the horizontal line:

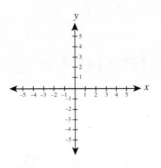

Notice that these lines intersect at 0. This is important.

These two lines together form a plane. Now you can use a single point to represent numbers on both lines. Suppose you want to represent the number 2 on the horizontal line, and the number 3 on the vertical line. Start by plotting those points on the respective lines:

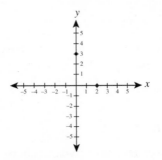

To find the point that represents both of the points we just plotted, we extend dotted lines from them and finding where they intersect:

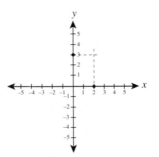

Now mark the place where the dotted lines intersect with a point, and you have just plotted a point on the coordinate plane.

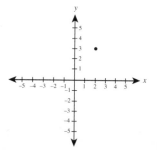

Now let's introduce a few more concepts. The horizontal and vertical lines are called *axes* (plural of *axis*). Every point on the plane corresponds to a point on each axis. The axes intersect at a point called the *origin*.

The horizontal axis is commonly referred to as the *x*–axis, and the vertical axis is called the *y*–axis. We specify a point's location with a pair of numbers in parentheses; the first number is the location along the *x*–axis, and the second number is the location along the *y*–axis. So we can give the coordinates of the point as (2, 3). Note that the coordinates (3, 2) pick out an entirely different point.

ORDERED PAIRS

Pairs of coordinates representing points on the plane are known as *ordered pairs*. This is because the order in which the numbers appear matters. (–4, 5) and (5, –4) pick out different points, even though they use the same numbers.

There are three important things to remember her:

1. The origin always has the coordinates (0, 0). (After all, the lines intersect each other at point 0.)
2. It is customary to represent a point with coordinates (*x*, *y*).
3. It is also customary to label the axes with *x* and *y*:

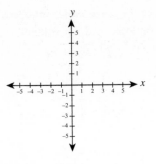

The Coordinate Grid

The coordinate plane is also known as the coordinate grid because it can be drawn with lines like those on a sheet of graph paper:

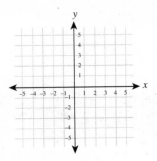

These lines will help you to plot points more precisely. That is especially true when you encounter a grid like this one, where the units are not labeled:

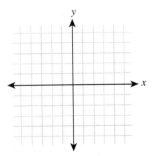

To plot points correctly here, you would need to count units from the origin on your own. Positive units are to the right of the origin on the *x*–axis, and above the origin on the *y*–axis. On the opposite sides of the origin, the negative numbers decrease as you go further away from it.

Lines, Line Segments, and Slope

Let's start by plotting two points on the coordinate grid: (2, 3) and (−2, −1):

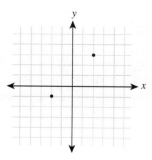

You will often find points labeled along with letters. Let's label the point at (2, 3) as *A*, and the point at (−2, −1) as *B*. Next, we'll connect the two points with a line:

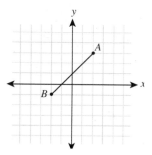

Strictly speaking, what connects *A* and *B* is a *line segment*. A *line* has no endpoints; it extends in both directions without end. A line segment, however, is a section of a line, and does have one or two endpoints. If it has one endpoint, it continues in a single direction and is called a ray.

Lines and line segments contain points. Every point on line segment *AB* is contained by it. Looking at the above graph, you might see that the points $(-1, 0)$, $(0, 1)$, and $(1, 2)$ are also on the line, as well as an infinite number of points whose coordinates can be expressed with fractions or decimals.

The Slope Formula

Lines and line segments have several important properties. One of them is *slope*. The slope of a line is its direction and "steepness." As well, the slope of a line is the ratio of the difference in *y*–coordinates to the difference between their *x*–coordinates.

Take two points on a line or line segment; call the coordinates of one point x_1 and y_1, and the coordinates of the other point x_2 and y_2. The slope of the line—represented by the letter *m*—is the value given by the following equation:

$$m = \frac{y_2 - y_1}{x_2 - x_1}$$

This equation is a *formula*. In algebra, a formula represents an important relationship in numbers and symbols. Not all of the equations we deal with are considered formulas, because they don't all represent important relationships.

One reason why we study coordinate geometry in Algebra I is that the slope of a line can be represented with a formula. The slope formula is but one important application of this.

To find the value of *m*, we simply substitute values for the variables on the right side of the equation and then evaluate the expression.

Let's find the slope of line segment *AB*. Which point represents x_1 and y_1 and which point represents x_2 and y_2 isn't important, as long as we stick to the assignment. So let's take the point *A* and make $x_1 = 2$ and $y_1 = 3$. Using the coordinates of *B*, we'll say $x_2 = -2$ and $y_2 = -1$.

$$m = \frac{-1 - 3}{-2 - 2} = \frac{-4}{-4} = 1$$

The slope of line segment *AB*, then, is 1.

A line's slope remains the same throughout. No matter what points on the line you use to calculate the slope, the results should be the same. You can check this for yourself by using the coordinates of any point that falls on *AB*.

One thing to note is that lines steeper than *AB* will have slopes greater than 1. Lines less steep will have slopes less than 1. A perfectly horizontal line has a slope of 0; since the difference between the *y*–coordinates of any two points on the line is 0, the numerator of the slope formula will be 0. A vertical line has an *undefined* slope. The denominator of the slope formula for a vertical line is 0, and such fractions are considered to have undefined values.

SLOPE AND DIRECTION

You'll notice that lines with positive slopes get higher as you move from left to right on the plane. Lines with negative slopes "head down" as you move from left to right. Keep this in mind and you'll be able to spot an incorrect slope value when you have a graph to look at.

The Midpoint Formula

Every line segment with two endpoints has a *midpoint*, the point halfway between the endpoints. (Note that this doesn't apply to line segments with only one endpoint, as a ray going in one direction has no middle!)

Let's call the coordinates of the midpoint of a line segment x_m and y_m. The midpoint is given by the formula:

$$\left(x_m, y_m\right) = \left(\frac{x_1 + x_2}{2}, \frac{y_1 + y_2}{2}\right)$$

Basically, the *x*–coordinate of the midpoint is halfway between the *x*–coordinates of the endpoints, and the *y*–coordinate of the midpoint is halfway between the *y*–coordinates of the endpoints.

Let's use our line segment AB as an example. Plugging the coordinates of A and B into the midpoint formula, we get:

$$\left(x_m, y_m\right) = \left(\frac{2 + -2}{2}, \frac{3 + -1}{2}\right) = \left(\frac{0}{2}, \frac{2}{2}\right) = (0, 1)$$

So the midpoint of AB is at (0, 1). Now let's visualize this:

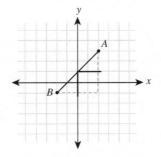

We draw horizontal and vertical lines that represent the horizontal and vertical distances between A and B. We then mark the midpoints on each line, and draw lines extending from them. The midpoint is the point where those lines intersect.

The Distance Formula

The distance formula equation allows us to use the coordinates of a line segment's two endpoints to calculate its length. This formula is based on the Pythagorean Theorem, which has to do with right triangles. The idea is that a line segment can usually be described as the *hypotenuse* of a right triangle. The hypotenuse is the longest side of a right triangle; it is always the one opposite the right angle.

In the coordinate grid above, the large triangle we graphed is a right triangle, and AB is the hypotenuse. Now, in any right triangle, the square of the length of the hypotenuse is equal to the sum of the squares of the lengths of the other two sides. So if C is the length of the hypotenuse and A and B are the lengths of the other sides, then:

$A^2 + B^2 = C^2$

The distance formula applies the Pythagorean Theorem to coordinate geometry. Where d is the length of the line segment with endpoints at coordinates (x_1, y_1) and (x_2, y_2),

$$d = \sqrt{(x_2 - x_1)^2 + (y_2 - y_1)^2}$$

The difference between the x–coordinates represents the length of the horizontal side of a right triangle. The difference between the y–coordinates represents the length of the vertical side.

STEPS YOU NEED TO REMEMBER

When plotting points and applying formulas, keeping track of your coordinates and the points they pick out is crucial.

Using *X*– and *Y*–Coordinates

Remember the following steps:

- Pick out the x–coordinate from the first number in the ordered pair of coordinates.
- Identify x–coordinates on the coordinate plane by tracking horizontally.
- Pick out the y–coordinate from the second number in the ordered pair of coordinates.
- Identify y–coordinates on the coordinate plane by tracking vertically.

We use the *subscripts* 1 and 2 to identify particular x– and y–coordinates when we are dealing with to points. So x_1 and y_1 are the coordinates of one point, and x_2 and y_2 are the coordinates of the other point. Be careful not to mix things up by taking x_1 and x_2 or x_1 and y_2 to be the coordinates of a particular point. Keep track of your coordinate and plug them in carefully.

Also, it does matter which point you assign coordinates x_1 and y_1. As long as you stick to your assignment, the results of applying the slope, midpoint, and distance formulas will be the same no matter which point gets subscript 1.

STEP–BY–STEP ILLUSTRATION OF THE 5 MOST COMMON QUESTION TYPES

The following five questions will walk you through the important steps involved in coordinate geometry. The most common questions involve using coordinates to work with points, lines, and line segments.

Question 1: Plotting Points

What point is three units above and two units to the left of the origin?

(A) $(-3, -2)$

(B) $(-2, 3)$

(C) $(2, 3)$

(D) $(3, 2)$

(E) $(3, -2)$

To identify the points described, trace the movement from the origin. The first arrow going from the origin goes to the point $(0, 3)$. This is three units *above* 0 on the vertical number line. We then move two units to the right; on the horizontal number line, that is -2. So the point has coordinates $(-2, 3)$, and **(B) is correct**.

Don't let the wording of the question throw you; even though the first number given is 3, it represents the *y*–coordinate. The *y*–coordinate is always the *second number in the pair*. Getting that order backward could lead to $(3, -2)$, choice (E), as an incorrect answer.

Remember, too, that the direction on the *x*–axis is important. If you thought that points to the right of the origin have positive *x*–coordinates, you might have picked $(2, 3)$ as the coordinates.

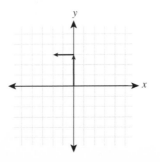

Question 2: Identifying Coordinates

What are the coordinates of the point of intersection of the two lines graphed below?

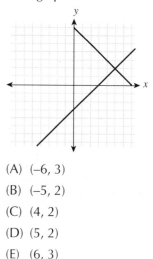

(A) (−6, 3)

(B) (−5, 2)

(C) (4, 2)

(D) (5, 2)

(E) (6, 3)

Here, you need only be concerned with the two lines' point of intersection. An intersection is used as a way to indirectly identify points in coordinate geometry. Since every point corresponds to numbers on the horizontal and vertical number lines, you can trace the point to each axis, as shown here.

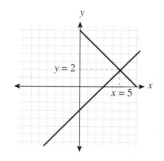

Since the point of intersection corresponds to 5 on the horizontal number line and 2 on the vertical number, the x–coordinate is 5 and the y–coordinate is 2. **So Choice (D) is the correct answer.** Choice (B), (−5, 2), might be tempting here for someone that visualizes the horizontal number line backwards. Some of the other choices like (4, 2) and (6, 3), simply represent errors in counting units from the origin.

Question 3: Finding the Slope

What is the slope of the line that passes through the points (–2, 4) and (–5, 10)?

(A) –3

(B) –2

(C) $-\dfrac{1}{2}$

(D) $\dfrac{1}{3}$

(E) 2

Once you have the coordinates for two points on a line, you can plug them into the slope formula we introduced earlier:

$$m = \frac{y_2 - y_1}{x_2 - x_1}$$

Let's say that $(x_1, y_1) = (-2, 4)$, and $(x_2, y_2) = (-5, 10)$. Plugging these values into the slope formula gets us:

$$m = \frac{10 - 4}{-5 - (-2)} = \frac{6}{-5 + 2} = \frac{6}{-3} = -2$$

So **(B) is the answer.** Remember that the *y*–coordinates get plugged into the numerator. If you plugged them into the denominator and used the *x*–coordinates in the numerator, we would have gotten a slope $-\frac{1}{2}$ choice (C). The value of –2 in (B) is what you would get by adding the coordinate values instead of subtracting. Keep in mind that we could have assigned the values of $x_1, y_1, x_2,$ and y_2 in the opposite way, and the result would still be the same.

Question 4: Finding the Midpoint

What is the midpoint of the line segment formed by the points $(-3, -4)$ and $(11, -6)$?

(A) $(-9, 7)$

(B) $(-3, 15)$

(C) $(4, -5)$

(D) $(8, -10)$

(E) $(14, -2)$

To find the coordinates of the midpoint, we plug the coordinates of the endpoint into the midpoint formula, which we presented earlier:

$$(x_m, y_m) = \left(\frac{x_1 + x_2}{2}, \frac{y_1 + y_2}{2} \right)$$

We can actually separate this into two formulas—one for each coordinate—and work them out one at a time.

$$x_m = \frac{x_1 + x_2}{2}$$

$$y_m = \frac{y_1 + y_2}{2}$$

Suppose $(x_1, y_1) = (-3, -4)$, and $(x_2, y_2) = (11, -6)$. Once again, we could make the opposite assignment, and the result would be the same. The important thing is to stick with an assignment of values once you make one. The next step is to plug in the values and then evaluate the resulting expressions.

$$x_m = \frac{-3 + 11}{2} = \frac{8}{2} = 4$$

$$y_m = \frac{-4 + -6}{2} = \frac{-10}{2} = -5$$

So the coordinates of the midpoint are $(4, -5)$, and (C) is correct. The coordinates $(8, -10)$ in (D) are simply the sums of the x– and y–coordinates the question provides. The coordinates $(14, -2)$ in (E) involves subtracting those coordinates.

Question 5: Calculating Distance

What is the length of the line segment *JK* graphed below?

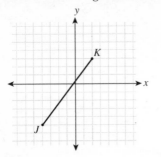

(A) 7

(B) 9

(C) 10

(D) 12

(E) 14

Here we need to use the distance formula:

$$d = \sqrt{(x_2 - x_1)^2 + (y_2 - y_1)^2}$$

But first, you must identify the coordinates of the points plotted on the coordinate grid.

　J: (−4, −5)

　K: (2, 3)

So let's say that $x_1 = -4$, $y_1 = -5$, $x_2, = 2$, and $y_2 = 3$. Now we have everything we need to find the length of the line segment. Let's plug in the coordinates:

$$
\begin{aligned}
d &= \sqrt{(2 - (-4))^2 + (3 - (-5))^2} \\
&= \sqrt{(2 + 4)^2 + (3 + 5)^2} \\
&= \sqrt{6^2 + 8^2} \\
&= \sqrt{36 + 64} \\
&= \sqrt{100} \\
&= 10
\end{aligned}
$$

Choice (C) is correct.

In working out the distance formula, we can see that *JK* is the hypotenuse of a right triangle with sides of lengths 6 and 8.

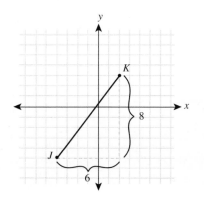

By subtracting coordinates, the distance formula uses those lengths to find the length of the third side.

The value of 7 in Choice (A) is what you would get by adding the side lengths and dividing by 2. Someone might get that as a result by confusing the midpoint formula with the distance formula. 14, Choice (E), is simply sum of the lengths of the two sides. In fact, the length of a hypotenuse is always less than the sum of the other side lengths.

CHAPTER QUIZ

1. Look at the three points graphed below.

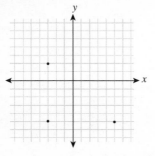

In order to graph the fourth corner of a rectangle, a point should be plotted at what coordinates?

(A) (2, 5)

(B) (3, 2)

(C) (4, 5)

(D) (5, 2)

(E) (5, 4)

2. What are the coordinates of the point *P* graphed below?

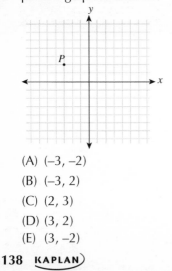

(A) (−3, −2)

(B) (−3, 2)

(C) (2, 3)

(D) (3, 2)

(E) (3, −2)

3. Which point graphed below has coordinates (3, −4)?

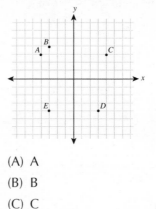

(A) A

(B) B

(C) C

(D) D

(E) E

4. Which point lies on the x–axis?

(A) (−2, −3)

(B) (0, 5)

(C) (1, 2)

(D) (3, 1)

(E) (4, 0)

5. What is the slope of the line passing through the two points graphed below?

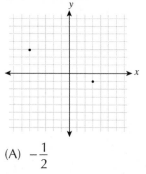

(A) $-\dfrac{1}{2}$

(B) $-\dfrac{1}{3}$

(C) $\dfrac{1}{3}$

(D) $\dfrac{1}{2}$

(E) 2

6. What is the slope of the line graphed below?

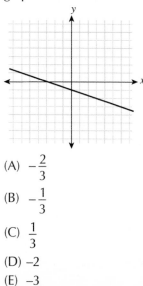

(A) $-\dfrac{2}{3}$

(B) $-\dfrac{1}{3}$

(C) $\dfrac{1}{3}$

(D) –2

(E) –3

7. A line segment has an endpoint at (–1, 3) and a midpoint at (5, 7). What are the coordinates of the other endpoint?

(A) (2, 5)

(B) (4, 8)

(C) (7, 15)

(D) (9, 12)

(E) (11,11)

8. What is the length of the line segment with endpoints at (5, 4) and (10, –8)?

(A) 9

(B) 11

(C) 13

(D) 17

(E) 19

ANSWER EXPLANATIONS

1. D

You need to plot the point that is the fourth corner and then identify its coordinates. Since this is a rectangle, the last corner must be in the upper left area of the plane. Let's plot the sides of the triangle to pinpoint the corner's coordinates.

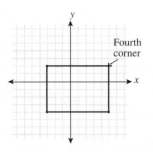

The fourth corner is where the top and right sides intersect. That point is five units to the left of the origin, and so the *x*–coordinate is 5. The point is two units above the origin, and so the *y*–coordinate is 2. The coordinates of the point are (5, 2). Choice (A), (2, 5), is tempting, because it has the coordinates reversed. (3, 2), B, is also tempting because it's the same distance from the *y*–axis as the corner located at (–3, 2).

2. B

The point is three units to the left of the origin, and two units above it. So the *x*–coordinate is –3 and the *y*–coordinate is 2. Since the *x*–coordinate comes before the *y*–coordinate, the location is (–3, 2).

The actual locations of the all of the coordinates listed in the answer choices are here:

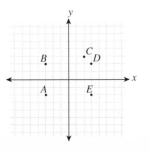

3. D

The point (3, –4) is 3 units to the right of 0 on the horizontal number line, and four units below 0 on the vertical number line. Let's trace a path:

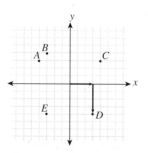

As you can see, the path leads directly to point *D*. All of the other points are three units above, below, left or right of the origin and four units away from it in another direction. Point B, for instance, has coordinates (–3, 4) and Point *C* has coordinates (4, 3).

4. E

The *x*–axis is the horizontal one that extends from the origin. The *y*–coordinate of every point on the *x*–axis is 0. The only point among the among the answer choices with such a *y*–coordinate is (4, 0).

Choice (B) is the most commonly selected incorrect answer, as those coordinates are for a point on the *y*–axis.

5. A

The points plotted above are located at coordinates (–5, 3) and (3, –1). So x_1 = –5, y_1 = 3, x_2, = 3, and y_2 = –1. Plug those values into the slope formula:

$$m = \frac{-1-3}{3-(-5)} = \frac{-4}{3+5} = -\frac{4}{8} = -\frac{1}{2}$$

You might have gotten 2 for the slope if you used the incorrect $m = \frac{x_1 - y_1}{x_2 - y_1}$

You might have gotten $\frac{1}{2}$ for the slope if you used the incorrect formula

$m = \frac{y_2 - y_1}{x_1 - x_2}$

6. B

When you're asked to find the slope of a graph like the one above, you need to supply the coordinates for points yourself. Pick any pair of coordinates that the line clearly goes through. The easiest thing to do is to use points on the x– and y– axes. It will make the slope formula calculations easier, since two of the four coordinates will be 0. The line intersects the x–axis at (–3, 0), and it intersects the y–axis at (0, –1). So let's say that $x_1 = -3$, $y_1 = 0$, x_2, $= 0$, and $y_2 = -1$:

$$m = \frac{y_2 - y_1}{x_2 - x_1} = \frac{-1-0}{0-(-3)} = -\frac{1}{3}$$

Choice (C), $\frac{1}{3}$ could be the result of not getting a positive number when subtracting –3 from 0. Choice (E), –3, is a common incorrect answer that one might get by using the inverse of the slope formula

7. E

This question involves the midpoint formula, but it is not a straightforward matter of plugging in numbers and evaluating. Here, you actually have to solve equations.

The midpoint formula is $(x_m, y_m) = \left(\frac{x_1 + x_2}{2}, \frac{y_1 + y_2}{2} \right)$

This actually involves two equations, one for the x–coordinates and one for the y–coordinates:

$$x_m = \frac{x_1 + x_2}{2}$$
$$y_m = \frac{y_1 + y_2}{2}$$

Let's plug in the coordinates we have into the equations. $(x_m, y_m) = (5, 7)$, and $(x_1, y_1) = (-1, 3)$. So

$$5 = \frac{-1 + x_2}{2}$$

$$7 = \frac{3 + y_2}{2}$$

Solve the x–coordinate equation first. Since the expression $(-1 + x_2)$ is divided by 2, we need to multiply both sides by 2:

$$5 \bullet 2 = \frac{-1 + x_2}{2} \bullet 2$$

$$10 = -1 + x_2$$

Next, subtract -1 from both sides to cancel out the addition on the right side. That is the same as adding 1:

$$10 = -1 + x_2$$
$$\underline{+1 \quad +1}$$
$$11 = x_2$$

So the x–coordinate of the other endpoint is 11. Now solve the other equation.

$$7 \bullet 2 = \frac{3 + y_2}{2} \bullet 2$$

$$14 = 3 + y_2$$

$$-3 \; -3$$

$$11 = y_2$$

So the y–coordinate of the other endpoint is also 11. The coordinates (2, 5) in Choice (A) actually represent the midpoint between the two points given in the question. The real endpoint is not in between the two points you are given. You might have gotten (7, 15) if you mismatched the x– and y– coordinates of the endpoint with the two equations.

8. C

This question requires the distance formula. We can use the coordinate values $x_1 = 5$, $x_2, = 10$, $y_1 = 4$, and $y_2 = -8$.

$$d = \sqrt{(10 - 5)^2 + (-8 - 4)^2}$$
$$= \sqrt{5^2 + (-12)^2}$$
$$= \sqrt{25 + 144}$$
$$= \sqrt{169}$$
$$= 13$$

You might have gotten 17, (D), if you had added the difference between the *x*–coordinates and difference between the *y*–coordinates. The Pythagorean Theorem does involve subtraction of that sort, but it also requires you to square the results.

CHAPTER 7

Graphing Linear Equations

WHAT ARE LINEAR EQUATIONS?

Different points on the plane can represent values of variables in an equation. Take the equation $y = 2x$. If $x = 2$, then $y = 4$. If $x = 3$, then $y = 6$. Each of these pairs of corresponding values for x and y can be plotted on the coordinate plane. In fact, all of the possible pairs of values for the variables would form a continuous line.

Let's plot the points we just derived, as well as a few others we could get from the equation $y = 2x$, and the draw a line through them.

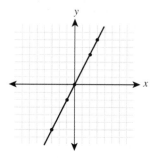

Every single point that lies on the line we just plotted "satisfies" the equation $y = 2x$. You would find that the value of the y–coordinate of any point is twice that of its x–coordinate.

$y = 2x$ is a *linear equation*. A linear equation is one whose graph is a straight line. In this chapter, our focus is on graphing linear equations. By using the properties of linear equations, we can graph them without having to plot many individual points.

CONCEPTS TO HELP YOU

By the end of this chapter, you should be able to interpret linear equations to get the information you need to graph them. The most reliable way to interpret them involves getting them into a standard form, called *slope–intercept form*. To use this slope–intercept form, you'll need to know about *x*– and *y*– intercepts.

X– and Y–Intercepts

A straight line graphed on the coordinate plane crosses at least one of the axes. The point where a line crosses an axis is called the *intercept*. A line crosses the *y*–axis at the *y*–intercept. The *x*–intercept of a line always has a *y*–coordinate of 0, and the *y*–intercept of a line always has an *x*–coordinate of 0. Since one coordinate of each intercept is already known, we can usually identify an intercept with a single number, rather than an ordered pair. If a line passes through the point $(7, 0)$, we could say that the line has an *x*–intercept of 7.

Linear equations can help to identify the intercepts of a line. In those cases, we can graph the equation of a line by plotting a straight line through both intercepts. In fact, all you ever need to graph a line is two points.

Suppose we know that a line has an *x*–intercept of –4 and a *y*–intercept of 2. That's already enough information to graph the line:

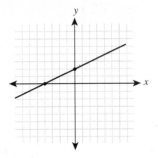

The two points plotted above are the intercepts. There is only one line that passes through both of them.

The Slope–Intercept Form

Using the slope formula from chapter 6, finding the slope of this line is straightforward. Let's take the coordinates of the intercepts of the line graphed above and plug them into the formula:

$$m = \frac{y_2 - y_1}{x_2 - x_1} = \frac{2 - 0}{0 - (-4)} = \frac{2}{4} = \frac{1}{2}$$

We've already talked about the slope of a line in terms of a relationship between the x– and y–coordinates of its points. The slope is the ratio of the change in y–coordinate values to the change in x–coordinate values between any two points on a line. So if a line has a slope of 3, it means that for every unit increase for x, the value of the y–coordinate increases by 3. Likewise, if a line has a slope of $\frac{1}{2}$, then for every unit increase for x, the value of the y–coordinate increases by $\frac{1}{2}$. That is, the y–coordinates increase at half the rate of the x–coordinates.

This relationship can be expressed with an equation. If $y = \frac{1}{2}x$, or $y = \frac{x}{2}$, then the value of y increases at half the rate at which x increases. What this means is that an equation of the form $y = mx$, describes a line with slope m.

Yet, while the line on the coordinate plane above has a slope of $\frac{1}{2}$, it's not hard to show that the line is not described by the equation $y = \frac{x}{2}$. If you look at the above graph, you can see that where the x–coordinate is 2, the y–coordinate is 3. Where the x–coordinate is 4, the y–coordinate is 4. So there's more to describing this line than the slope. Fortunately, there is a way to account for this extra factor in linear equations.

Note that the y–intercept is 2. When $x = 0$, $y = 2$. Since $\frac{x}{2} = 0$ when $x = 0$, you could get a value of 2 by adding that number to $\frac{x}{2}$. So let's take $\frac{x}{2} + 2$ and evaluate it for several values of x. If $x = 4$, $\frac{x}{2} + 2 = 4$. If $x = -2$, $\frac{x}{2} + 2 = 1$. If $x = -4$, $\frac{x}{2} + 2 = 0$. If you look at the graph again, you'll see that these values match the y–coordinates for each value of x.

Indeed, this line has a slope of $\frac{1}{2}$, and each y–coordinate can be obtained by adding 2 to half of the value of the x–coordinate. So this line can be described by the following equation:

$$y = \frac{x}{2} + 2$$

This is a linear equation that appears in a very common and useful form: the *slope–intercept form*. This is the form $y = mx + b$, where m is the slope and b is the y–intercept of the line.

Why does b always turns out to be the y–intercept? Since the x–coordinate at the y–intercept is 0, if $x = 0$, then $y = mx + b = m(0) + b = 0 + b = b$. So where the x–coordinate is 0, the y–coordinate is b.

STEPS YOU NEED TO REMEMBER

Though an equation doesn't strictly need to be in slope–intercept form for graphing purposes, it certainly helps. Having a standard form of an equation makes everything easier.

Getting Equations into Slope–Intercept Form

Any linear equation can be converted to slope–intercept form by solving for y. We already covered solving two–variable equations for one variable in chapter 5. For good measure, write the equation using anything added or subtracted to the number as a separate term, and not as part of a fraction including the variable.

The following equation is in slope–intercept form:

$$y = \frac{3x}{2} + 4$$

The following equations are *not* in slope–intercept form:

$$2y = 3x + 8$$
$$y = \frac{3x + 8}{2}$$
$$x = \frac{2x}{3} - 4$$

$$y - \frac{3x}{2} = 4$$

$$y - 4 = \frac{3x}{2}$$

NON-LINEAR EQUATIONS

Many equations have graphs that are not straight lines. Quadratic equations, for instance, have U-shaped graphs, so they are not linear equations.

If the solution of an equation for y has any of the following features, then you are not dealing with a linear equation:

• x is raised to a power

• x is inside a radical sign

• x is in the denominator of a fraction

If the solution of an equation meets any of these conditions, then you would find that it has a curved graph. When that is the case, you will not be able to get the equation into slope–intercept form.

Using the Slope–Intercept Form to Get the Intercepts

Once you have an equation in slope–intercept form, finding the intercepts is straightforward. You can read the y–intercept right off the equation, but getting the x–intercept will require a bit more work.

Let's take our earlier equation, $y = \frac{3x}{2} + 4$. In the form $y = mx + b$, $b = 4$, so the y–intercept is 4. Now, to find the x–intercept, you need to solve the equation for x when $y = 0$. The idea is that you need to find the x–coordinate of the point on the line where the y–coordinate is 0. So if $y = 0$,

$$\frac{3x}{2} + 4 = 0$$

Let's solve this equation for x. First, subtract 4 from both sides to get

$$\frac{3x}{2} = -4$$

Now, since the variable x is multiplied by $\frac{3}{2}$, you need to divide both sides by that number. That amounts to multiplying both sides by $\frac{2}{3}$, the multiplicative inverse of $\frac{3}{2}$:

$$\frac{3x}{2} \cdot \frac{2}{3} = -4 \cdot \frac{2}{3}$$

$$x = -\frac{8}{3}$$

And so the *x*–intercept of the line is $-\frac{8}{3}$.

Once you have both intercepts of a line, you can graph the equation by plotting those two points and drawing a line through them.

CALCULATING THE *X*–INTERCEPT

There is a shortcut to the process for finding the *x*–intercept. Once you have the linear equation in the form $y = mx + b$, the *x*–intercept is $-\frac{b}{m}$. Notice that we actually just solved the equation for *x* when $y = 0$ by dividing $-b$ by *m*.

STEP–BY–STEP ILLUSTRATION OF THE 5 MOST COMMON QUESTION TYPES

The following questions address the process of graphing linear equations, as well as the methods for using graphs to get equations.

Question 1: Getting Equations into Slope–Intercept Form

What is the slope–intercept form of $10x = 3y - 2x - 6$?

(A) $y = \frac{8x}{3} - 2$

(B) $y = \frac{8x}{3} + 2$

(C) $y = \frac{8x}{3} + 6$

(D) $y = 4x + 2$

(E) $y = 4x + 6$

Getting an equation into slope intercept form is a matter of solving for y, so we need to isolate y. Following the order of operations, we begin by adding 6 to both sides:

$$10x = 3y - 2x - 6$$
$$\underline{+6 \qquad\qquad +6}$$
$$10x + 6 = 3y - 2x \qquad\qquad \text{Now add } 2x \text{ to both sides}$$

$$10x + 6 = 3y - 2x$$
$$\underline{+2x \qquad\qquad +2x}$$
$$12x + 6 = 3y \qquad\qquad \text{Divide both sides by 3 since } y \text{ is}$$
$$\text{multiplied by that number:}$$

$$12x + 6 = 3y$$
$$\div 3 \qquad\quad \div 3$$
$$4x + 2 = y \qquad\qquad \text{Now put the equation into the}$$
$$y = mx + b \text{ form}$$

$$y = 4x + 2$$

Choice (D) is the answer. One might arrive at (B), $y = \frac{8x}{3} + 2$ by subtracting $2x$ from $10x$ instead of adding. (E) might seem correct if you only carried out division on $12x$ when dividing $12x + 6$ by 3.

Question 2: Using Coordinates to Derive Linear Equations

What is the equation of the line that passes through the points $(1, -4)$ and $(7, 8)$?

(A) $y = \dfrac{x}{2} - \dfrac{9}{2}$

(B) $y = \dfrac{x}{2} + \dfrac{9}{2}$

(C) $y = 2x - 9$

(D) $y = 2x - 7$

(E) $y = 2x - 6$

To find the equation of line in slope–intercept form using the coordinates of two points, you should begin by finding the slope. Here, $x_1 = 1$, $x_2 = -4$,

$y_1 = -4$, and $y_2 = 8$. According the slope formula,

$$m = \frac{y_2 - y_1}{x_2 - x_1} = \frac{8 - (-4)}{7 - 1} = \frac{12}{6} = 2$$ So the slope of the line is 2. Now rewrite the equation

$$y = 2x + b$$

The problem here is that we don't yet have a value for b, the y–intercept. The key to getting that value is solving the equation for b, using values for x and y. Since we were give the coordinates for two points, we can use one of those ordered pairs as values for x and y, respectively. Let's take the point (7, 8):

$x = 7$ and $y = 8$ Since $y = 2x + b$, we can rewrite the equation

$8 = 2(7) + b$

$8 = 14 + b$

$b = -6$ Now we can see the the complete equation for the line

$y = 2x - 6$

Choice (E) is correct. Note that you would get the same result if you used the coordinates (1, –4).

Question 3: Finding Intercepts to Graph Equations

What are the intercepts of $y = 4x - 1$?

(A) x–intercept: –1; y–intercept: $\frac{1}{4}$

(B) x–intercept: $-\frac{1}{4}$; y–intercept: –4

(C) x–intercept: $-\frac{1}{4}$; y–intercept: –1

(D) x–intercept: $\frac{1}{4}$; y–intercept: –1

(E) x–intercept: $\frac{1}{4}$; y–intercept: 4

Since b is the y–intercept in the equation $y = mx + b$, that value can be read directly off the equation. The y–intercept is –1. As we learned earlier, the x–intercept of a line $y = mx + b$ is $-\frac{b}{m}$. In $y = 4x - 1$, $m = 4$ and $b = -1$. So $-\frac{b}{m} = -\frac{-1}{4} = \frac{1}{4}$, and so the x–intercept is $\frac{1}{4}$.

(D) is the correct answer. You would have instead gotten $-\frac{1}{4}$ for the x–intercept, as in (B) and (C), if you used the $\frac{b}{m}$ to get the value instead. (E) has the correct x–intercept, by mistakenly uses the value of the slope for the y–intercept.

Question 4: Graphing Absolute Value Equations

What is the graph of $y = |x - 1|$?

The graph of a linear absolute value equation involves parts of two straight lines. Since y is equal to an absolute value in this equation, y cannot have a negative value,—no matter what the value of x may be. In fact, whenever $x - 1$ is less than 0, $y = -(x - 1)$. That's how absolute values work, after all.

So in order to graph, $y = |x - 1|$, we have to graph two lines: $y = x - 1$ and $y = -(x - 1)$. Distributing the negative sign in the second equation gets us a slope–intercept form equation $y = -x + 1$.

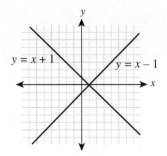

The graph of $y = |x - 1|$ is made up of the line segments above the x–axis. Anything below that axis is not part of the graph, because y–coordinates can have only positive values. So this is the graph of the absolute value equation:

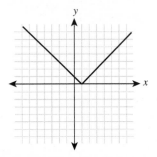

This graph matches the one in **choice (A), and so (A) is correct**. Notice that for every x–coordinate less than 1, $x - 1$ would be less than 0. For those x–coordinates, you take the negative of the value of $x - 1$ for the y–coordinate.

The graph in (B) is actually that of the equation $y = |x| + 1$. The graph in (C) includes the line segment corresponding to $y = x - 1$, but not the part corresponding to $y = -x + 1$. As long as the equation gives us y–coordinates for points where x is less than 0, those points must be graphed. The graph in (D) is actually that of the equation $x = |y| + 1$.

Question 5: Graphing Equations With 0 or Undefined Slopes

Which of these is the graph of a line with an undefined slope?

(A)   (B)   (C)

(D)   (E)

A fraction is undefined if its denominator equals 0. Since the slope of the line is given by the formula $m = \frac{y_2 - y_1}{x_2 - x_1}$, the slope is undefined if x_1 and x_2 have the same value. Since the x–coordinate of each point on a vertical line is the same, such lines have undefined slopes. For any two points on a vertical line, $x_1 = x_2$, and so $m = \frac{y_2 - y_1}{0}$. m will be undefined, no matter what values of y_1 and y_2 we work with. The vertical line in (E) has an undefined slope, so **(E) is correct**.

Keep in mind that an undefined value is not the same thing as 0. For horizontal lines like the one in (B), every y–coordinate has the same value. Since $y_2 - y_1 = 0$ if $y_1 = y_2$, $m = \frac{0}{x_2 - x_1} = 0$. Every horizontal line on the coordinate plane has a slope of 0.

CHAPTER QUIZ

1. The slope–intercept form of the equation $4x - 3y + 5 = 0$ is

   (A) $x = \dfrac{3}{4}y - \dfrac{5}{4}$

   (B) $x = -\dfrac{3}{4}x - \dfrac{5}{4}$

   (C) $y = \dfrac{3}{4}x - \dfrac{3}{5}$

   (D) $y = \dfrac{3}{4}x - \dfrac{5}{4}$

   (E) $y = \dfrac{4x}{3} + \dfrac{5}{3}$

2. Which of the following is the graph of $2y + 3x = 8$?

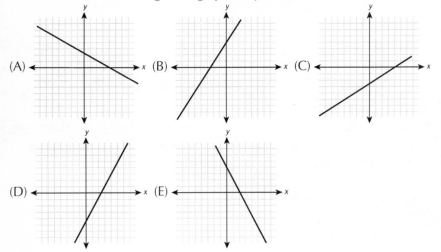

3. The equation of the line shown below is

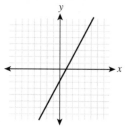

(A) $y = -2x + 2$

(B) $y = \dfrac{x}{2} + 2$

(C) $y = \dfrac{x}{2} + 1$

(D) $y = 2x - 2$

(E) $y = 2x + 1$

4. The equation of a line with a slope of 3 and an *x*–intercept of –2 is

(A) $y = -2x + \dfrac{2}{3}$

(B) $y = -2x + 3$

(C) $y = 3x - 2$

(D) $y = 3x + 6$

(E) $y = 3x + \dfrac{2}{3}$

5. The equation of the line that passes through the point (–3, –2) and (12, 4) is

(A) $y = \dfrac{5x}{2} - \dfrac{11}{2}$

(B) $y = \dfrac{5x}{2} + \dfrac{11}{2}$

(C) $y = \dfrac{2x}{5} + \dfrac{1}{5}$

(D) $y = \dfrac{2x}{5} - \dfrac{1}{5}$

(E) $y = \dfrac{2x}{5} - \dfrac{4}{5}$

6. A line has an x–intercept of 3 and a y–intercept of –5. What is the equation of the line?

(A) $y = \frac{3}{5}x - 5$

(B) $y = \frac{3}{5}x + 3$

(C) $y = \frac{5}{3}x - 5$

(D) $y = \frac{5}{3}x + 3$

(E) $y = \frac{5}{3}x + 5$

7. Which of the following is the graph of $y = |2x + 1|$?

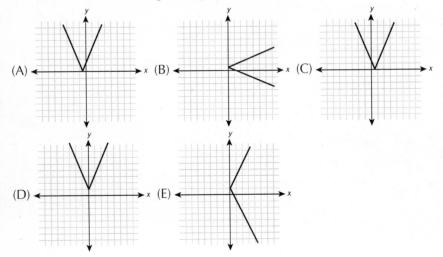

8. Which line has a slope of 0?

(A) $x = -1$

(B) $y = 2$

(C) $x = 3$

(D) $y = x$

(E) $y = \frac{1}{x}$

ANSWER EXPLANATIONS

1. E

You need to solve for y.

$$4x - 3y + 5 = 0$$
$$\underline{+ 3y \qquad +3y}$$
$$4x + 5 = 3y$$
$$\div 3 \qquad \div 3$$
$$\frac{4x + 5}{3} = y$$
$$y = \frac{4x}{3} + \frac{5}{3}$$

Choice (A) is actually the solution of the equation for x, and (D) is the solution for y of $3x - 4y + 5 = 0$.

2. E

The first step in graphing this equation is getting it in slope–intercept form:

$$2y + 3x = 8$$
$$\underline{-3x \qquad -3x}$$
$$2y = -3x + 8$$
$$\div 2 \qquad \div 2$$
$$y = -\frac{3x}{2} + 4$$

So the slope m is $-\frac{3}{2}$, the y–intercept b is 4, and the x–intercept $-\frac{b}{m}$ is $\frac{8}{3}$.

So the line $2y + 3x = 8$ passes through the points (0, 4), and ($\frac{8}{3}$, 0), which are plotted here:

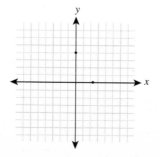

Only the graph in (E) includes both of these points. Choice (A) mixes up the x– and y–intercepts, and (B) has the wrong x–intercept ($-\frac{8}{3}$ instead of $\frac{8}{3}$). That could be the result of getting an incorrect slope of $-\frac{3}{2}$.

3. D

The line has an x–intercept of 1 and a y–intercept of –2. So we can use the slope formula, where, $x_1 = 0$, $x_2 = 1$, $y_1 = -2$, and $y_2 = 0$. According the slope formula,

$$m = \frac{y_2 - y_1}{x_2 - x_1} = \frac{0-(-2)}{1-0} = \frac{2}{1} = 2$$

So $m = 2$ and $b = -2$. That makes the slope–intercept form of the linear equation $y = 2x - 2$. (C) and (E) mistakenly use of the value of the x–intercept for b. (B) and (C) calculate the slope using the formula

$$m = \frac{x_2 - x_1}{y_2 - y_1} \text{ instead of } m = \frac{y_2 - y_1}{x_2 - x_1}.$$

4. D

To find the right equation, we need the value of the y–intercept in addition to the slope, which is provided. Since the value of the x–intercept is equal to $-\frac{b}{m}$, we can find the value of the y–intercept, b, by solving the equation $-2 = -\frac{b}{m}$ for b. Since the slope is 3, $m = 3$, and $-2 = -\frac{b}{3}$. We can solve the equation by multiplying both sides by –3. That leaves us with $b = 6$. Since the slope–intercept form is $y = mx + b$, the equation of our line is $y = 3x + 6$. One might choose $y = 3x - 2$ as a result of confusing the x– and y–intercepts. $y = -2x + 3$, Choice (B), is the result of using the incorrect equation $y = bx + m$. Choice (E) could be result of solving $0 = 3x - 2$ for x and using the solution as the y–intercept.

5. E

We can use the coordinates provided to find the slope of the line.

$$m = \frac{y_2 - y_1}{x_2 - x_1} = \frac{4-(-2)}{12-(-3)} = \frac{4+2}{12+3} = \frac{6}{15} = \frac{2}{5}$$

So the equation of the line is $y = \frac{2x}{5} + b$. We need to find the value of b. Plug in one pair of coordinates, and then solve the resulting equation for b.

$$-2 = \frac{2(-3)}{5} + b$$

$$-2 = -\frac{6}{5} + b$$

$$b = -2 + \frac{6}{5} = -\frac{10}{5} + \frac{6}{5} = -\frac{4}{5}$$

So the equation of the line is $y = \frac{2x}{5} - \frac{4}{5}$ and (E) is correct. (A) and (B) involve using the inverse of the slope formula. The y–intercept of $\frac{11}{2}$ in (B) would be found by plugging in the coordinates $(-3, -2)$ in $y = \frac{5x}{2} + b$ and solving the equation for b.

6. C

We can take the coordinates of the intercepts and plug them into the slope formula.
$$m = \frac{y_2 - y_1}{x_2 - x_1} = \frac{0 - (-5)}{3 - 0} = \frac{5}{3}$$

Since the y–intercept is -5, $b = -5$. $m = \frac{5}{3}$, and so the equation is $y = \frac{5}{3}x - 5$.

(A) and (B) involve the use of the incorrect slope formula $m = \frac{x_2 - x_1}{y_2 - y_1}$.

7. A

Graphing this absolute value equation involves using the graphs of both $y = 2x + 1$ and $y = -2x - 1$ ($y = -(2x - 1)$). Both lines are graphed here:

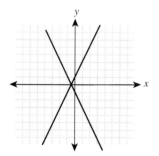

Since we are graphing an absolute value equation, only the points with positive y–coordinates appear. This means that the graph matching the line segments above the x–axis represents the equation $y = |2x + 1|$. The graph in (A) is a match. The graph in (C), however, is that of $y = |2x - 1|$, and the graph in (D) is that of $y = |2x| + 1$.

8. B

A line has slope 0 if all of the y–coordinates are the same. Only if $y_1 = y_2$ will $m = \dfrac{y_2 - y_1}{x_2 - x_1} = \dfrac{0}{x_2 - x_1} = 0$. The equation $y = 2$ describes the line where each point has a y–coordinate of 2. Since all of the y–coordinates are the same on that line, it has a slope of 0. The lines $x = -1$ and $x = 3$ are both vertical lines with undefined slopes.

Systems of Equations

WHAT ARE SYSTEMS OF EQUATIONS?

In explained in chapter 5, you cannot solve a single equation for two variables; the most you can do is solve for one variable in terms of the other one. However, when given a *system of equations* (a group of two or more equations), you can find the values of both variables. This chapter will introduce several ways to work with systems of equations.

CONCEPTS TO HELP YOU

Two approaches in particular can be of help in solving systems of equations. Both help you to obtain a single–variable equation which can then be solved.

Combining Equations

It is possible to take two equations, each with two variables, and combine them to make a new equation with a single variable. Look at this system of equations:

$$10 = x + 3$$
$$10 = 3x - 11$$

When dealing with a system of equations, each instance of a variable always stands for the same number. So whatever the value of x may turn out to be, it should be the same in both equations.

Now, the equations tell us that both $x + 3$ and $3x - 11$ have the value of 10. Since those expressions have the same value,

$$x + 3 = 3x - 11$$

You could solve this equation to find that $x = 7$. Then again, you could have done that with either of the equations we started with. We combined equations here just to give you an example. When it comes to equations with two variables, however, combination can be the key to the solution.

Suppose you are given these two linear equations:

$$y = 4x - 7$$
$$y = 3x - 2$$

Since each of the expressions on the right side equals y, they are equal to each other:

$$4x - 7 = 3x - 2$$

This is step 1 in our first approach to solving systems of equations. We'll go through the entire process a little later in the chapter.

Substitution

A second approach to solving systems of equations involves substitution. Substitution is the process of replacing a variable in an equation with a equivalent expression. The idea is that when you have one expression solved for one variable, you can substitute its value into the other equation.

$$a = 4b + 5$$
$$2a + 3b = 8b$$

Since a and $4b + 5$ have the same value, you can substitute any instance of a with $4b + 5$.

$$2a + 3b = 8b$$
$$2(4b + 5) + 3b = 8b$$
$$8b + 10 + 3b = 8b$$

At this point, finding the value of b is straightforward.

COMBINING EQUATIONS

Strictly speaking, combining equations can be thought of as substitution. It is a matter of substituting one instance of a variable with an equivalent expression. Combination is a special case of substitution, where both equations are already solved in terms of one variable.

STEPS YOU NEED TO REMEMBER

Solving a system of equations can involve a series of steps: First, you use combination or substitution to get the value of one variable. Second, you use the value of the first variable to find the value of the second variable.

Getting and Solving Single–Variable Equations

The first step in solving a system of equations is to get a single variable equation. Depending on the system you are working with, you can choose to use combination or substitution.

If both equations are already stated in terms of solutions for the same variable (such as two linear equations in slope–intercept form), then *combination* is the way to go. You can also use combination by solving both of the equations of the same variable. If that's what combination would require, however, you might be better off using substitution. Then you would need only to solve one of the equations for one variable, and then substitute the solution into the other equation.

Once you have a single variable equation, you can solve it as you would any equation.

Using the Single–Variable Equation Solution

Now we come to a new step. Once you have found the value of one variable in a two–variable system, you need to go back to the other equation you were initially given (the one you didn't use to solve for the first variable). Plug in the value of the variable you obtained in the previous step, and you'll have an equation with one variable that should be easy to solve.

STEP–BY–STEP ILLUSTRATION OF THE 5 MOST COMMON QUESTION TYPES

Now we'll use combination and substitution to solve systems of equations, including a system of three equations. We'll also see how solving systems of equations is used in coordinate geometry and in evaluating expressions.

Keep in mind that there are likely to be several paths you can take to get to the solution of a system of equations. Some of the answer explanations below might not match the approach you would take on your own.

Question 1: Combining Variables

In the system of equations below, what is the value of a?

$a = 4b + 9$

$a = 7b - 3$

(A) 2

(B) 4

(C) 11

(D) 17

(E) 25

Since both of the equations here have the same variable alone on one side, you could say that you already have two solutions of a in terms of b. This system, then, is tailor–made for solving by combination. Since both $4b + 9$ and $7b - 3$ equal a,

$4b + 9 = 7b - 3$ Add 3 to both sides and then subtract $4b$ from both sides

$12 = 3b$ Divide both sides by 3

$b = 4$

Now, we have the value of one variable. But we're not done; we need to find the value of a. We need to plug our value of b into one of the equations.

$$4b + 9$$
$$a = 4(4) + 9$$
$$= 16 + 9$$
$$= 25$$

Choice (E) is correct. Choice (B), 4 gives the value of b, not a. (A), 2, is the value of b had you subtracted, not added, 3 from both sides of $4b + 9 = 7b - 3$.

Question 2: Solving by Substitution

If $3x - 18 = 2y + 21$ and $4x = y + 32$, what are the values of x and y?

(A) $x = 3$ and $y = -9$

(B) $x = 3$ and $y = -20$

(C) $x = 5$ and $y = -12$

(D) $x = 5$ and $y = -9$

(E) $x = 5$ and $y = 9$

You could use the combination process to answer this question, but that would require solving both solutions for either x or y. It's simpler to solve one equation for one variable, and then substitute the solution into the other equation. Since solving the equation $4x = y + 32$ for y requires only one step, let's go with that.

Subtracting 32 from both sides of $4x = y + 32$ gives us $y = 4x - 32$. Now we can substitute $4x - 32$ for y in $3x - 18 = 2y + 21$:

$$3x - 18 = 2(4x - 32) + 21$$
$$3x - 18 = 8x - 64 + 21$$

$3x - 18 = 8x - 43$	Add 43 to both sides, then subtract $3x$ from both sides
$25 = 5x$	Divide both sides by 5
$x = 5$	

Now let's plug 5 back into $4x = y + 32$ and solve for y:

$$4(5) = y + 32$$
$$20 = y + 32$$
$$y = -12$$

So **(C) is the correct answer.** Choices (A) and (B) might have resulted from getting 15 instead of 25 when adding 43 to −18. The values for y in those choices each result from plugging 3 into the different equations and solving. (D) and (E) have the wrong y–coordinates, perhaps the result of plugging 5 into the longer equation $3x − 18 = 2y + 21$. The longer the equation, the greater the likelihood of making an error in the solution process.

Question 3: Intersections of Lines

At what point do the lines $y = 2x + 10$ and $y = 6x + 18$ intersect?

(A) $(-7, -4)$

(B) $(-2, 6)$

(C) $(-1, 8)$

(D) $(1, 12)$

(E) $(2, 14)$

Though this might seem like a graphing question, it can be answered by solving a system of equations. The lines intersect at the point that lies on both of them. Therefore, we are looking for a pair of coordinates (values for x and y) that fit both equations. So you can find the x– and y–coordinates by solving the system of equations.

Since each equation gives the value of y in terms of x, we can use the combination process from Question 1 above. Were one or both of the equations not in slope–intercept form, the substitution process would be more attractive. Since $2x + 10$ and $6x + 18$ both equal y,

$2x + 10 = 6x + 18$	Subtract 10 from both sides and then subtract $6x$ from both sides
$-4x = 8$	Divide both sides by -4
$x = -2$	

So we now know the x–coordinate of the point of intersection. To get the y–coordinate, we plug the value of x into one of the equations.

$$y = 2x + 10$$
$$y = 2(-2) + 10$$
$$= -4 + 10$$
$$= 6$$

The coordinates of the point are $(-2, 6)$, and so **(B) is correct**. You can check this by plugging in the values of the x– and y–coordinates for x and y in each equation and finding they hold.

The coordinates in choice (A), $(-7, -4)$, could have been obtained by adding 10 to the right side instead of subtracting in solving the equation $2x + 10 = 6x + 18$, and then plugging the result into $y = 2x + 10$ to get the y–coordinate. The coordinates in (D) could be the result of adding $6x$ to the left side instead of subtracting in solving the equation $2x + 10 = 6x + 18$, and then plugging the result into $y = 2x + 10$ to get the y–coordinate.

Question 4: Solving and Evaluating

If $4s - 11 = 35 - 2t$ and $8 - 6s = 2 - 4t$, then $2s + t =$

(A) 23

(B) 25

(C) 27

(D) 30

(E) 32

This question involves a little more than solving a system of equations; it requires you to use the solution to evaluate another expression. You might be tempted to combine the two equations in a way that provides a value for $2s + t$, but that would prove to be frustrating. Your best bet is to get a solution for the system of equations as you did in the previous questions and then use those values with the expression.

Let's use substitution to solve this system. We can start by solving the equation for $4s - 11 = 35 - 2t$. You could solve either expression for either variable, but this one looks like it has a solution with no fractions. If you can avoid working with fractions, so much the better.

$$4s - 11 = 35 - 2t \qquad \text{Subtract 35 from both sides}$$
$$4s - 46 = -2t \qquad \text{Divide both sides by } -2$$
$$t = 23 - 2s$$

Now that one equation is solved for t, we can substitute that value t in the other equation:

$$8 - 6s = 2 - 4(23 - 2s)$$
$$8 - 6s = 2 - 92 - (-8s)$$
$$8 - 6s = 8s - 90 \qquad \text{Subtract on both sides}$$
$$14s = 98 \qquad \qquad \text{Divide both sides by 14 to get the value of } s$$
$$s = 7$$

The next step is to plug that value into one of the questions, and solve for t. Since $s = 7$:

$$8 - 6(7) = 2 - 4t$$
$$8 - 42 = 2 - 4t$$
$$-34 = 2 - 4t$$
$$4t = 36$$
$$t = 9$$
$$s = 7$$
$$t = 9$$

So $2s + t = 2(7) + 9 = 14 + 9 = 23$ and **(A) is correct.** Choice (D), 30, is the result of getting a value of 9 for s and plugging in that value in $8 - 6s = 2 - 4t$. That would get you a value of 12 for t.

Question 5: Solving Systems of Three Equations

Looking at the system of equations below, what is the value of t?

$$s - 3r = t - 1$$
$$2s - t = 4r + 2$$
$$2r + 2t = 5 - s$$

(A) −8

(B) −4

(C) 6

(D) 9

(E) 16

This question features a system including three variables. When three variables are involved, you'll need at least three equations. Since solving a three–equation system can be long and complicated, be ready to carry out a process of trial and error. You might find it helpful to use both combination and substitution. You can involve only two equations in a particular step, but all three equations must be involved over the entire process.

As always, we take steps with an eye toward getting new equations with fewer variables.

Let's try combination first, by solving two equations for t.

$$s - 3r = t - 1$$
$$s - 3r + 1 = t$$
$$2s - t = 4r + 2$$
$$2s - 4r - 2 = t$$

So $s - 3r + 1 = 2s - 4r - 2$ Solve for s

$$r + 3 = s$$

Now, let's substitute the value of s into the equation $s - 3r = t - 1$, and solve the resulting equation for t:

$$(r + 3) - 3r = t - 1$$
$$-2r + 3 = t - 1$$
$$t = -2r + 4$$

Let's do the same thing with the equation $2r + 2t = 5 - s$:

$$2r + 2t = 5 - (r + 3)$$
$$2r + 2t = 5 - r - 3$$
$$2t = 2 - 3r$$
$$t = 1 - \frac{3}{2}$$

Since $t = -2r + 4$ and $t = 1 - \frac{3r}{2}$:

$$-2r + 4 = 1 - \frac{3r}{2}$$
$$\frac{r}{2} = 3$$
$$r = 6$$

Since $t = -2r + 4$:

$$t = -2(6) + 4 = -12 + 4 = -8$$

(A) is the correct answer. Choices (C), 6, and (D), 9, are the values of r and s, respectively.

CHAPTER QUIZ

1. If $p = r - 7$ and $p = \frac{r}{2} + 5$, what is the value of p?

2. Looking at the pair of equations below, what is the value of b?

$$3a - 2b = 4$$
$$b = 2a - 6$$

   (A) -22
   (B) -8
   (C) 8
   (D) 10
   (E) 26

3. If $5j + 3k = 84$ and $7k - 11j = -8$, what are the values of j and k?

   (A) $j = 6$ and $k = 38$
   (B) $j = 6$ and $k = 18$
   (C) $j = 9$ and $k = 13$
   (D) $j = 9$ and $k = 20$
   (E) $j = 9$ and $k = 33$

4. If $2c + 3d = 25$ and $4c + 2d = 22$, then $c + d =$

   (A) 6
   (B) 7
   (C) 8
   (D) 9
   (E) 10

5. If $4 - 2g = 14 - 2h$ and $5g - 4h = -31$, then $h - 2g =$

   (A)   1

   (B)   6

   (C)   8

   (D) 12

   (E) 16

6. What is the point of intersection of $y = 3x - 5$ and $y = 2x + 3$

   (A) $(-8, -13)$

   (B) $(-2, -11)$

   (C) $(-2, -1)$

   (D) $(8, 17)$

   (E) $(8, 19)$

7. Looking at the system of equations below, what is the value of z?

$$3x + 2 = 11 - 2y$$
$$2x = 1 - 7z$$
$$2y = 9z + 3$$

   (A) -12

   (B) -3

   (C)   5

   (D)   8

   (E)   10

8. Looking at the system of equations below, what is the value of $a + b - c$?

$$a - b = c + 2$$
$$2b + a = c - 10$$
$$a + c = -2b$$

   (A) -6

   (B) -4

   (C)   2

   (D)   4

   (E) 12

ANSWER EXPLANATIONS

1. 17

It's easiest to do combination here. In both equations, the value of p is expressed in terms of the r.

$$r - 7 = \frac{r}{2} + 5 \qquad \text{Add 7 to both sides and subtract } \frac{r}{2} \text{ from both sides}$$

$$\frac{r}{2} = 12 \qquad \text{Multiply both sides by 2 to get the value of } r$$

$$r = 24 \qquad \text{Now plug this value in to one of equations, } p = r - 7$$

$$p = 24 - 7 = 17$$

Keep in mind the specific variable you're asked to give a value for. Many people would report 24 as the value of p, when it is actually the value of r. Another common error involves getting a value of 6 instead of 24 for r, as a result of dividing 12 by 2 instead of multiplying. That in turn could result in a value of -1 or 8 for p, depending on which equation you plug in the incorrect value of r.

2. D

Since one equation gives the value of b in terms of a, the easiest approach here is substitution. Let's substitute b with $2a - 6$ in $3a - 2b = 4$, and solve for a:

$$3a - 2(2a - 6) = 4$$
$$3a - 4a + 12 = 4$$
$$-a + 12 = 4$$
$$a = 8$$

Since $b = 2a - 6$:

$$b = 2(8) - 6 = 16 - 6 = 10$$

Choice (A), -22, could be the result of getting -8 instead of 8 for the value of a, and then using $b = 2a - 6$ to get the value of b. (C), 8, reports the correct value of the wrong variable. (E), 26, is what you might get if you found a to have a value of 16 instead of 8.

3.   C

This question lends itself to substitution. Solving both equations for one variable would involve too much work. Instead, let's solve one equation, $5j + 3k = 84$, for k:

$$5j + 3k = 84$$
$$3k = 84 - 5j$$
$$p = \frac{r}{2} + 5$$

Now let's substitute this value of k in the other equation:

$$7\left(28 - \frac{5j}{3}\right) - 11j = -8$$
$$196 - \frac{35j}{3} - 11j = -8$$
$$\frac{35j}{3} + 11j = 204$$
$$\frac{35j}{3} + \frac{33j}{3} = 204$$
$$\frac{68j}{3} = 204$$
$$68j = 612$$
$$j = 9$$

To find the value of k, replace j with 9 in one of the equations, and solve:

$$5(9) + 3k = 84$$
$$45 + 3k = 84$$
$$3k = 39$$
$$k = 13$$

(A) and (B) result from getting 6 instead of 9 as the value of j in $68j = 612$. Plugging that value into $5j + 3k = 84$ would get you $k = 18$, (B). You would get 38, (A), if you added 30 to 84 instead of subtracting when solving $30 + 3k = 84$. (E) has the correct value of 9 for j, but the value of 33 for k is incorrect; perhaps the result of adding 45 to 84 instead of subtracting when solving $45 + 3k = 84$.

4. D

This question asks you evaluate an expression with two variables. You'll need to solve the system of equations in order to get those variables. We'll carry out substitution here, using the solution of $4c + 2d = 22$ for d.

$$4c + 2d = 22$$
$$2d = 22 - 4c$$
$$d = 11 - 2c$$

Now replace d with $11 - 2c$ in $2c + 3d = 25$:

$$2c + 3(11 - 2c) = 25$$
$$2c + 33 - 6c = 25$$
$$4c = 8$$
$$c = 2$$

Since $4c + 2d = 22$ and $c = 2$,

$$8 + 2d = 22$$
$$2d = 14$$
$$d = 7$$

So $c + d = 2 + 7 = 9$.

You might have gotten a value of 8 if you got $-8c = 8$ instead of $4c = 8$ during the solution process. A value of -1 for c would get you a value of 9 for d, as a solution of $4c + 2d = 22$. If you instead got $8c = 8$, you would get a value of 10 for $c + d$.

5. E

Let's begin by solving $4 - 2g = 14 - 2h$ for one variable. That looks to be a better option than solving the other equation for one variable, as that would involve fractions.

Since $4 - 2g = 14 - 2h$:

$$2h = 2g + 10$$
$$h = g + 5 \qquad \text{Use that value to solve } 5g - 4h = -31$$
$$5g - 4(g + 5) = -31$$
$$5g - 4g - 20 = -31$$
$$g = -11$$
$$h = g + 5 = -11 + 5 = -6$$

Therefore, $h - 2g = -6 - 2(-11) = -6 + 22 = 16$. Choice (B), 6, would result from getting a value of -16 instead of -6 for h.

6. E

The coordinates of the point of intersection are the pair values of x and y that is the solution to the system of linear equations. Since we are dealing with linear equations in slope–intercept form, combination is the way to go. Both $3x - 5$ and $2x + 3$ equal y:

$$3x - 5 = 2x + 3$$
$$3x = 2x + 8$$
$$x = 8$$

Since $y = 3x - 5$:

$$y = 3(8) - 5 = 24 - 5 = 19$$

So the point of intersection is $(8, 19)$. One might get an incorrect x–coordinate of -2 by subtract adding -5 instead of 5 to 3 when solving $3x - 5 = 2x + 3$. In that case, you could get -11 or -1 for the y–coordinate, depending on which equation you solve for y when $x = -2$.

7. B

This system of the equations is a bit different from the one in question 5. Here, each equation involves only two of the three variables. We'll get started by combining equations, and we'll take a shortcut.

Since $3x + 2 = 11 - 2y$, $2y = 9 - 3x$. Now we have two equations that give the value of two y, and we can combine them:

$$2y = 9 - 3x$$

So $9 - 3x = 9z + 3$ Solve that equation for x

$$x = 2 - 3z$$ Now plug that value into $2x = 1 - 7z$:

$$2(2 - 3z) = 1 - 7z$$
$$4 - 6z = 1 - 7z$$
$$z = -3$$

The value of y in this system is -12, and the value of x is 8. (C), 5, is what you would get if you added 4 to $1 - 7z$ instead of subtracting while solving $4 - 6z = 1 - 7z$.

8.   A

Begin by solving two equations for a in terms of b and c.

$$a - b = c + 2$$
$$a = b + c + 2$$
$$2b + a = c - 10$$
$$a = c - 10 - 2b$$

$$b + c + 2 = c - 10 - 2b \qquad \text{Subtract } c \text{ from both sides, eliminating it}$$
$$\text{altogether}$$

$$3b = -12$$
$$b = -4$$

Now plug the value of b into $a - b = c + 2$:

$$a - -4 = c + 2$$
$$a = c - 2$$

Taking $a + c = -2b$, we find that:

$$(c - 2) + c = -2(-4)$$
$$2c - 2 = 8$$
$$2c = 10$$
$$c = 5$$
$$a = c - 2 = 5 - 2 = 3$$
$$a + b - c = 3 + (-4) - 5 = -1 - 5 = -6$$

(B), -4, would result from getting 1 instead of -1 when combining 3 and -4. Choice (C), 2, is actually the value of $a - b - c$. (D), 4, is the value of $a + b + c$. (E), 12, is the value of $a - b + c$.

CHAPTER 9

Linear and Compound Inequalities

WHAT ARE LINEAR AND COMPOUND INEQUALITIES?

While equations are mathematical statements that relate expressions as equal, *inequalities* relate statements as being greater than or less than another. A *linear inequality* is one whose graph is an area of the coordinate plane that is bounded by a straight line. A *compound inequality* combines two or more linear inequalities with "and" or "or." The graph of a compound inequality is one or more areas of the coordinate plane, bounded by at least two straight lines.

The keystone of the equation is the 'equals' sign you're familiar with. Inequalities, on the other hand, can feature different signs; expressions are related as *less than*, *greater than*, *less than* "or" *equal to*, or *greater than* "or" *equal to*.

CONCEPTS TO HELP YOU

Our focus with linear inequalities is the same as our focus with linear equations: solving and graphing. Much of what we learned about equations can be applied to inequalities, but there are different ways to represent the solutions of inequalities.

Using the Symbols

Look at the following chart that explains the four inequality signs:

Sign	Example	Meaning
<	$4 < 7$	"Four is less than seven"
>	$8 > 5$	"Eight is greater than five"
≤	$x \leq 6$	"x is less than or equal to six"; "x is at most six"; "x is no more than six"
≥	$y \geq 9$	"y is greater than or equal to nine"; "y is at least nine"; "y is no less than nine"

REVERSING INEQUALITIES

Any inequality can be rewritten with the opposite sign, if you switch the expressions around. Since 3 is greater than 2, 2 is less than 3. Likewise, if $x \leq 5$, then $5 \geq x$.

Solutions of Linear Inequalities

A linear inequality can be solved in a manner similar to that of solving an equation.

We can write the solution of $2x + 5 = 21$ as $x = 8$.

Likewise, the solution of $2x + 5 < 21$ is $x < 8$, and the solution of $2x + 5 > 21$ is $x > 8$.

Using operations to solve an inequality can get a bit more complicated, as we'll see shortly.

The solution of a basic linear equation is a single number. Other equations we've looked at had two solutions. Linear inequalities, on the other hand, have a range of solutions. Remember, the solution of a linear equation is a value for the variable that makes the equation true.

When it comes to linear inequalities, however, there are many values the variable could have to make the inequality true. For instance, $x > 7$ is true if $x = 7.25$, and it is true if $x = 8$ or $x = 100$. Therefore, we usually state the solution of an inequality in terms of another inequality, as that is the best way to express a range of numbers.

Solutions and the Number Line

We introduced the number line in chapter 1 in order to explain certain number properties. It also happens than the number line is useful for representing the solutions to linear equalities visually.

To indicate the solution of an inequality, you can mark every point on the number line that makes the inequality true. In the case of a linear inequality, this is a matter of drawing a line segment on the number line.

Since the inequality $3x + 5 > 11$ for any value of x greater than 2, we can represent the solution, $x > 2$, on the number line:

Every point to the right of 2 is highlighted. You would find that $3x + 5 > 11$ is true for any of them, but not for any point to the left of 2. Notice that the point we laid down at 2 on the line is a hollow circle. There is a reason it looks like this: that the circle is hollow means that while every point to the right of 2 is a solution, 2 is *not* a solution. If $x = 2$, $3x + 5 = 11$, and 11 is not greater than 11! On the other hand, 2 is a solution to $3x + 5 \geq 11$. So we the solution to $3x + 5 \geq 11$ appears like this:

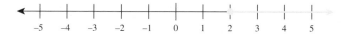

We use a solid circle instead of a hollow one to indicate that the number it highlights is part of the solution.

Compound Inequalities

Suppose we want to state that the expression $x + 8$ has a value between 5 and 10. We cannot do so with a single inequality sign. Essentially, we are saying that $x + 8 > 5$ and $x + 8 < 10$. This is a compound inequality, a combination of two more linear inequalities. We can combine linear inequalities with either "and" or "or." When they are combined with "and," (as in $x + 8 > 5$ and $x + 8 < 10$), they can be rewritten into a single inequality with two signs:

$$5 < x + 8 < 10$$

To solve this inequality, however, it is best to break it down into two inequalities.

$$x + 8 > 5 = x > -3$$
$$x + 8 < 10 = x < 2$$

The solutions can then be combined into a single inequality, $-3 < x < 2$. That solution can be represented on the number line as a line segment with two endpoints:

Suppose we are told instead that $x + 8 < 5$ "or" $x + 8 > 10$. This amounts to $x < -3$ "or" $x > 2$. Notice, first, that we could not express a compound inequality by combining these inequalities with "and." No number is both less than -3 and greater than 2! We can say that a number is one or the other, and represent that on the number line like this:

Since the value of x is not "between" two numbers, we can't express the solution—or the original compound inequality, for that matter—as a single inequality. That can only be done with those connecting inequalities with "and."

Graphing Inequalities

The number line can also be used to represent the solution of an inequality with one variable. If the linear equality involves two variables, however, you need the coordinate plane to represent it visually.

Let's take the linear inequality $2y - 2 > 4x$. The graph of this isn't a line, but it involves one. The solution of this inequality for y is $y > 2x + 1$. Take the inequality, and replace the inequality sign with the = sign to get $y = 2x + 1$. The graph of that line is:

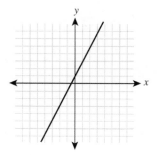

Since $y > 2x + 1$ is the solution to our inequality, any point with coordinates that satisfy $y > 2x + 1$ is part of the solution. Take the point $(-2, 3)$. When $x = -2$ and $y = 3$, y is greater than $2x + 1$. So that point is part of the graph of the solution.

In fact, all of the points "to the left" of the line we graphed have y–coordinates that are greater than $2x + 1$ (where x is the value of the x–coordinate). We indicate that by *shading*; the entire area of the coordinate plane that contains the solution should be shaded:

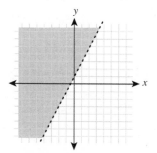

Notice that the original line we graphed is dotted. This indicates that the points on the line are not part of the solution, just as the hollow circle indicates that the point on the number line isn't part of the solution. Points on the line, such as (0, 1), do not satisfy $y > 2x + 1$, since $y = 1$ and $2x + 1 = 1$ at that point.

If, on the other hand, we were graphing the inequality $2y - 2 \geq 4x$, we would use a solid line:

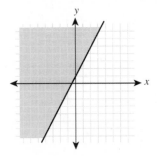

Since $1 = 1$, we can also say that $1 \geq 1$, and so the point (1, 1), like all of the points on the line $y = 2x + 1$, is part of the solution.

We'll talk about graphing compound inequalities in the next section.

STEPS YOU NEED TO REMEMBER

We will once again highlight some key ways in which solving and graphing inequalities departs from the steps we take in dealing with linear equation.

Multiplying and Dividing by Negative Numbers

In chapter 5, we explained the multiplication property of equality:

If $x = 10$, then $x * -5 = 10 * -5$.

Now consider the inequality $10 > 6$. That's certainly true, but what if we multiplied both sides by -5? You might expect the result to be $-50 > -30$, but that inequality is false! -50 is less than -30.

When you multiply (or divide) both sides of an inequality by a negative number, you **must change the direction of the inequality sign**. If you remember that step, the resulting inequality will be true.

Breaking Up and Recombining Compound Inequalities

To solve a compound inequality that joins linear inequalities with "and," break it up into two inequalities. One can include the "less than" sign, and the other can include "greater than."

$x - 4 < 3x + 6 < 45$

$x - 4 < 3x + 6$

$3x + 6 < 45$

After solving both inequalities (for the same variable), recombine them into a single inequality with two signs.

$-5 < x$

$x < 13$ Since x is greater than -5 and less than 13, rewrite as follows

$-5 < x < 13$

As long as the range of possible values of the variable falls between the numbers or expressions on the other sides of the inequalities, we can combine both solutions into one inequality.

Now, you might have solved $x - 4 < 3x + 6$ in a slightly different way so as to get $x > -5$. Remember that you can always turn an inequality around, switching the terms and reversing the sign. If x is greater than -5, then -5 is less than x.

Also, you may be able to combine linear inequalities even when the original one was not given as a single inequality with two signs. Some test questions are designed so that you will have to figure out whether the solutions you come up with can be combined.

Graphing Linear and Compound Inequalities

We began discussing the steps from graphing linear inequalities earlier in this chapter. The complete steps are as follows:

1. Replace the inequality sign with the equals sign and graph the resulting linear equation.

2. If the inequality has a > or < sign, make the line dotted. If it has a ≥ or a ≤ sign, make the line solid.

3. Shade the region of the graph on the side of the graph with the points whose coordinate values satisfy the inequality. If you graphed the line correctly, all of the points on one side of it will satisfy the inequality, and none of the points on the other side will satisfy it. To figure which side, just pick any point that is not directly on the line. If it satisfies the solution of the inequality, shade that side. If not, shade the other side.

How you graph a compound inequality depends in part on whether it is an "and" or an "or" inequality. In either case, though, you should begin by graphing both lines. Without actually shading anything, figure out which section of the plane would be shaded as the solution of each part of the inequality. If you have an "and" inequality, the shaded region is the area of *overlap* of those sections. If you have an "or" inequality, the shaded region is the *combination* of those sections.

STEP–BY–STEP ILLUSTRATION OF THE 5 MOST COMMON QUESTION TYPES

Our focus here is on solving linear and compound inequalities. Of course, a solution can be represented in different ways; it can take the form of another inequality, or it can be represented graphically, as a section of the coordinate plane or the number line.

Question 1: Solving a Linear Inequality

$8x + 24 < 2x - 12 =$

(A) $x < -6$

(B) $x > -2$

(C) $x < 2$

(D) $x < 6$

(E) $x > 6$

Solving a linear equality is much like solving an equation. The goal is to get the variable on one side of the equation.

$8x + 24 < 2x - 12$	Subtract 24 from both sides
$8x < 2x - 36$	Subtract $2x$ from both sides
$6x < -36$	Divide both sides by 6
$x < -6$	

Choice (A) is the answer. Choice (B), $x > -2$, is what you might select if you got $8x < 2x - 12$ instead of $8x < 2x - 36$ (you would have to reverse the direction of the inequality sign if you divided by a negative number). (E), $x > 6$, is the result of $-6x < -36$ instead of $6x < -36$.

Question 2: Solving a Compound Inequality

What is the solution of the compound inequality $2a - 1 < 5a + 8$ and $-4a + 19 > a - 6$?

(A) $-5 < a < 3$

(B) $-3 < a < 5$

(C) $5 < a < -3$

(D) $a > 3$ or $a < -5$

(E) $a < -3$ or $a > 5$

Here we must deal with each inequality one at a time.

$2a - 1 < 5a + 8$	Subtract 8 from both sides
$2a - 9 < 5a$	Subtract $2a$ from both sides
$-9 < 3a$	Divide both sides by 3
$-3 < a$	

In general, try to add or subtract so as to get a variable term with a positive value on one side of the inequality. Otherwise, you'll have to divide both sides by a negative number. That would happen if you went about solving this inequality a little differently, so as to get $-3a < 9$ instead of $-9 < 3a$. That process is more complicated, since you would need to change the direction of the inequality sign,

$-4a + 19 > a - 6$	Add 6 to both sides
$-4a + 25 > a$	Add $4a$ to both sides
$25 > 5a$	Divide both sides by 5
$5 > a$ or $a < 5$	

So a is greater than -3 and less than 5. Since the value is between -3 and 5, we can combine the inequalities into a single one with two signs, $-3 < a < 5$. **(B) is the answer**. (A), $-5 < a < 3$, is the solution to $2a + 1 > 5a - 8$ and $4a + 19 > -a - 6$. Choice (C), $5 < a < -3$. has the inequality signs reversed, so as to get an inequality that could not ever be true.

Question 3: Using the Number Line

Which number line shows the solution to the compound inequality $7 - 2x > 13$ or $1 - x \leq -3$?

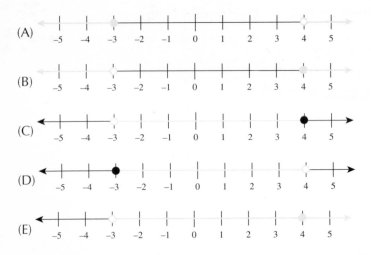

Start by solving both parts of this inequality.

$$7 - 2x > 13$$
$$-2x < 6$$
$$x < -3 \qquad \text{Change the direction of the inequality sign}$$
$$\text{when dividing by a negative number}$$
$$1 - x \leq -3$$
$$-x \leq -4$$
$$x \geq 4$$

So $x < -3$ "or" $x \geq 4$.

The first part of this solution is represented by every point to the left of -3 on the number line. Since we're using a "less than" sign, the solution does not include -3, and we mark that point with a hollow circle:

The second part of the solution is represented by every point to the right of 4 on the number line. Since we're using a "less than" sign, the solution includes 4, and so we mark that point with a solid circle:

The complete representation of the solution is the combination of these number lines, and so **(B) is the correct answer.** Choice (A) could be the result of misreading the inequality signs, as that solution is $x \leq -3$ "or" $x > 4$. The line in choice (E) shows the solution to $x > -3$ "and" $x \geq 4$. That actually amounts to just $x > -3$, since any number greater than or equal 4 is also greater than -3.

Question 4: Solving Absolute Value Inequalities

$|a| > 4 = ?$

(A) $a < -4$

(B) $a > -4$

(C) $-a < 4 < a$

(D) $a > 4$ or $a < -4$

(E) $-4 < a < 4$

Though it may not look like it at first glance, $|a| > 4$ is actually a compound inequality.

If a is greater than 0, then $a > 4$.

But if a is less than 0, then $-a > 4$.   Recall that $|a| = -a$ if $a < 0$)

So $a > 4$ (if a is positive), and $-a > 4$ (if a is negative). If you divide both sides of $-a > 4$ by -1, you get $a < -4$. So $a > 4$ or $a < -4$, and **(D) is correct.**

Choice (A), $a < -4$, only includes one part of the compound inequality. (C), $-a < 4 < a$, would be the correct solution of $|a| < 4$.

Question 5: Graphing Compound Inequalities

The graph of $36 - 18x \le 6y + 6 < 12x - 18$ is

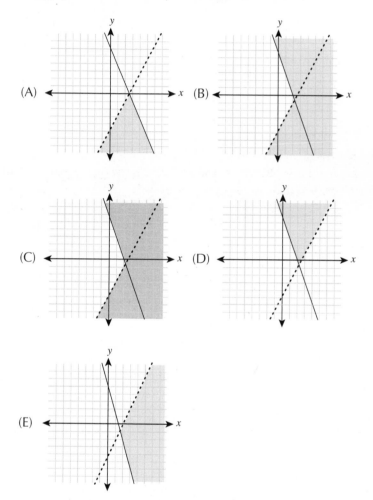

(A)

(B)

(C)

(D)

(E)

Let's first break this inequality into two linear inequalities:

$36 - 18x \le 6y + 6$

$y \ge -3x + 5$

$6y + 6 < 12x - 18$

$y < 2x - 4$

Next, let's graph the lines $y = -3x + 5$ and $y = 2x - 4$.

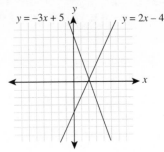

Now, any lines based on inequalities with "greater than or equal to" or "less than or equal to" signs should remain solid. Lines based on inequalities with "greater than" or "less than" signs should be dotted:

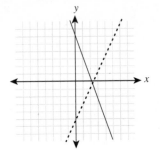

Now it is time to figure out which region of the graph should be shaded. Let's pick a single point, say, (3, 3). That point is on the left side of $y = 2x - 4$, but its coordinates do not satisfy $y < 2x - 4$, since $3 > 2(3) - 4$. So the graph of $y < 2x - 4$ is the section of the plane to the right of $y = 2x - 4$:

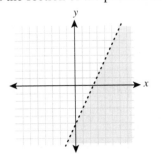

The point (3, 3) is on the right side of $y = -3x + 5$, and its coordinates satisfy $y \geq -3x + 5$, as $3 \geq -3(3) + 5$. So the graph of $y \geq -3x + 5$ is the section of the plane to the right of $y = -3x + 5$:

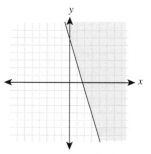

Since we are working with a compound "and" inequality, the graph is the area of *overlap* of these sections:

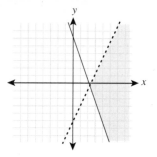

(E) is the answer. Choice (A) is the graph of $y \leq -3x + 5$ and $y < 2x - 4$. Choice (B) is the graph of $y < 2x - 4$ "or" $y \geq -3x + 5$, as it includes the combined graphs of $y < 2x - 4$ and $y \geq -3x + 5$, rather than the area of overlap.

CHAPTER QUIZ

1. $2x + 17 < 10x + 15 =$

   (A) $x < -4$

   (B) $x > -4$

   (C) $x > \dfrac{1}{4}$

   (D) $x > -\dfrac{1}{4}$

   (E) $x > 4$

2. $-\dfrac{5x}{4} > -\dfrac{3}{2} =$

   (A) $x > -\dfrac{15}{8}$

   (B) $x < -\dfrac{6}{5}$

   (C) $x > -\dfrac{6}{5}$

   (D) $x < \dfrac{6}{5}$

   (E) $x < \dfrac{15}{8}$

3. Which number line gives the solution of $8x + 12 < 12x - 4$?

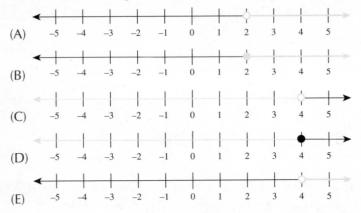

   (A)

   (B)

   (C)

   (D)

   (E)

4. The solution of $5a + 3 < b < 7a - 6$ for a is

   (A) $\dfrac{b}{7} + 6 < a < \dfrac{b}{5} - 3$

   (B) $\dfrac{b + 6}{7} < a < \dfrac{b - 3}{5}$

   (C) $\dfrac{b + 6}{7} > a$ or $a < \dfrac{b - 3}{5}$

   (D) $\dfrac{b + 6}{7} < a$ or $a > \dfrac{b - 3}{5}$

   (E) $\dfrac{b}{7} + 6 < a$ or $a > \dfrac{b}{5} - 3$

5. Which number is part of the solution of $3x - 1 < 5x + 11 < 2x + 62$?

   (A) −18

   (B) −6

   (C) 12

   (D) 18

   (E) 24

6. Which number line gives the solution of $|2x - 1| \le 5$

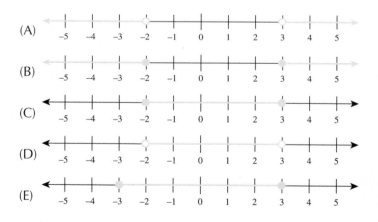

7. The lines $y = x - 2$ and $y = -\dfrac{2x}{3} + \dfrac{4}{3}$ divide the coordinate plane into four parts, I, II, III, and IV, as shown below.

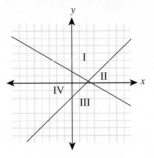

The graph of $x - 2 \leq y \leq -\dfrac{2x}{3} + \dfrac{4}{3}$ includes what section of the coordinate plane?

(A) I only

(B) IV only

(C) I and III

(D) II and IV

(E) I, III, and IV

8. Which of the following is the graph of $y \geq |2x - 1|$

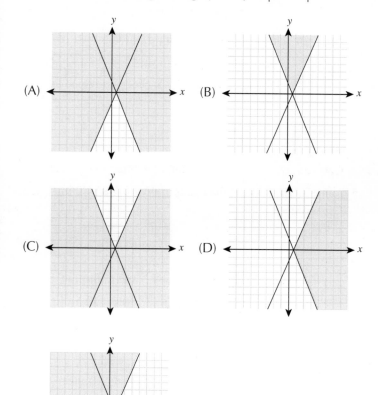

(A)

(B)

(C)

(D)

(E)

ANSWER EXPLANATIONS

1. **C**

   $2x + 17 < 10x + 15$ Subtract 15 from both sides

   $2x + 2 < 10x$ Subtract $2x$ from both sides

   $2 < 8x$ Divide both sides by 8. Since 2 divided by 8 is $\frac{1}{4}$:

   $\frac{1}{4} < x$

   $x > \frac{1}{4}$

   (C) is correct. Choice (D), $x > -\frac{1}{4}$, is the result of subtracting 17 from 15 instead of 15 from 17. (E), $x > 4$, is the result of dividing 8 by 2 instead of 2 by 8.

2. **D**

   $-\dfrac{5x}{4} > -\dfrac{3}{2}$ Divide both sides by $-\frac{5}{4}$ and multiply the right side, $-\frac{3}{2}$, by the multiplicative inverse of $-\frac{5}{4}$, which is $-\frac{4}{5}$.

   $-\dfrac{4}{5} \bullet -\dfrac{3}{2} = \dfrac{12}{10} = \dfrac{6}{5}$

   $x < \dfrac{6}{5}$

   (D) is correct. (B), $x < -\frac{6}{5}$, is what you get if you took the product of two negative numbers to be negative. (E) results from multiplying $-\frac{3}{2}$ by $-\frac{5}{4}$ instead of $-\frac{4}{5}$.

3. **E**

   $8x + 12 < 12x - 4$ Add 4 to both sides

   $8x + 16 < 12x$ Subtract $8x$ from both sides

   $16 < 4x$ Divide both sides by 4

   $4 < x$

So we mark off every point on the number line to the right of 4, but not 4 itself:

The number line in Choice (A) results from subtracting 4 from 12 instead of adding. That would lead to a solution of $x > 2$. The line in (B) represents $x \geq 2$, and the line in (C) is the solution of $8x + 12 > 12x - 4$.

4.   B

This inequality gives a solution in for y. We need to solve it for x, such that x is alone on one side of an inequality sign. We need to solve $5a + 3 < b$ and $b < 7a - 6$ separately.

$$5a + 3 < b$$

$$a < \frac{b-3}{5}$$

$$b < 7a - 6$$

$$\frac{b+6}{7} < a$$

$$\frac{b+6}{7} < a < \frac{b-3}{5}$$

(C), $\frac{b+6}{7} > a$ or $a < \frac{b-3}{5}$, has $\frac{b+6}{7} > a$ instead of $\frac{b+6}{7} < a$ for the solution of $b < 7a - 6$. (D) has $a > \frac{b-3}{5}$ instead of $a < \frac{b-3}{5}$ for the solution of $5a + 3 < b$.

5.   C

This question is a bit unique. We need to identify a number that satisfies the inequality. We'll solve it, and then see which number makes the solution true. This is easier than plugging each choice into each expression and evaluating. –6 would be part of the solution of the inequality

$3x - 1 \leq 5x + 11 < 2x + 62$, however.

$3x - 1 < 5x + 11 < 2x + 62$

$-6 < x$

$5x + 11 < 2x + 62$

$x < 17$

So $-6 < x < 17$.

Of all the numbers in the answer choices, only 12 is between -6 and 17, so (C) is correct. Choice (A), -18, would seem correct if you got $x < -6$ instead of $-6 < x$ as the solution of $3x - 1 < 5x + 11$. (B), -6, may be tempting because that number appears in the solution. However, since the inequality says that x is greater than -6, it can equal -6.

6. C

If $|2x - 1| \leq 5$, then $2x - 1 \leq 5$ (if $2x - 1$ is positive) and $-(2x - 1) \leq 5$ (if $2x - 1$ is negative). $x \leq 3$ (if $2x - 1$ is positive) and $x \geq -2$ (if $2x - 1$ is negative). We can combine those results to get $-2 \leq x \leq 3$

On the number line, that solution is the set of points between -2 and 3. We also include the numbers -2 and 3 by marking them with solid circles, since the inequality uses the "less than or equal to" sign. The result is

The number line in Choice (A) shows the solution to $|2x - 1| > 5$, while the line in (B) shows the solution to $|2x - 1| \geq 5$. The line in (D) involves the correct solution of the absolute value inequality, but it uses hollow circles, which signify "less than" and "greater than." (E) gets the solution $x \leq 3$ correct, but then uses -3 in the other solution instead of solving $-(2x - 1) \leq 5$.

7. E

To identify the part of the grid the graph of $x - 2 \le y \le -\frac{2x}{3} + \frac{4}{3}$ includes, you need to break this compound inequality up into two linear inequalities, and graph both them. We can rewrite $x - 2 \le y \le -\frac{2x}{3} + \frac{4}{3}$ as the pair $y \ge x - 2$ and $y \le -\frac{2x}{3} + \frac{4}{3}$.

The graph of $y \ge x - 2$ is

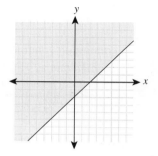

The graph of $y \le -\frac{2x}{3} + \frac{4}{3}$ is

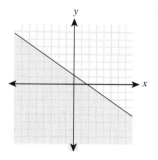

The graph of $x - 2 \le y \le -\frac{2x}{3} + \frac{4}{3}$, is the area of overlap of these two graphs. That is the region of IV.

Section I is the graph of $y \ge x - 2$ and $y \ge -\frac{2x}{3} + \frac{4}{3}$ .represent. Sections I, III, and IV make up the graph of $y \ge x - 2$ "or" $y \le -\frac{2x}{3} + \frac{4}{3}$.

8. B

Choice (A) is actually the graph of $y \geq -2x + 1$ or $y \geq 2x - 1$. That graph is the result of taking the absolute value inequality to be an "or" inequality instead of an "and inequality."

Since $|2x - 1|$ is $2x - 1$ if $2x - 1 > 0$, and it is $-(2x - 1)$ if $2x - 1 < 0$, we must graph the inequalities $y \geq 2x - 1$ and $y \geq -2x + 1$. We can start by plotting the lines $y = 2x - 1$ and $y = -2x + 1$:

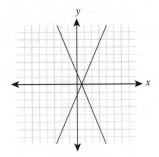

The graph of $y \geq 2x - 1$ is the region including the line $y = 2x - 1$, and every point to the left of it. If we take the point $(1, 5)$ and plug the x- and y-coordinate values into $y \geq 2x - 1$, we find that $5 \geq 2(1) - 1$ is true. The graph of $y \geq -2x + 1$ is the region including the line $y = -2x + 1$, and every point to the right of it. The same point $(1, 5)$ also satisfies that inequality. The graph of the compound inequality, then, is the area of overlap of the graphs of the linear inequalities:

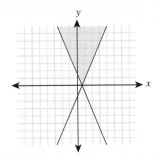

Choice (A) represents the graph of $y \geq -2x + 1$ or $y \geq 2x - 1$ (the result of taking the absolute value inequality to be an "or" compound inequality instead of an "and" inequality).

CHAPTER 10

Polynomials

WHAT ARE POLYNOMIALS?

Polynomials are best understood in terms of a related idea: *monomials*. A monomial is a number, a variable, or the product of a number and one or more variables. All of the following are monomials:

$$5$$
$$x$$
$$3y$$
$$2ab$$
$$4st^2$$
$$x^2 y^7 z^4$$

Expressions with addition or subtraction operations are not monomials.

Some of the above monomials such as $3y$ and $2ab$ have numbers as factors. We'll call those numbers *coefficients*.

A *polynomial* is an expression consisting of one or more monomials. If a polynomial contains more than one monomial, the monomials are combined with addition or subtraction. A *binomial* is a polynomial with two monomial terms:

$$5 + x$$
$$x + 2xy$$
$$ab - 4$$
$$y + y^2$$

A *trinomial* is a polynomial with three monomial terms:

$$x^2 - 4x + 5$$
$$a + ab + b$$
$$3x^6 + 2x^3 + 7x$$

A polynomial can have any number of terms, but monomials, binomials, and trinomials are common in Algebra I.

As you can see from the above examples, polynomials commonly include powers. We sometimes use these powers to classify polynomials by their *order*:

- $5x^3 + x^2 + x$ is a *third order* polynomial because the largest exponent is 3.

- $7x^6 + 2x^4 + 2x^3 + 7x$ is a *sixth order* polynomial because the largest exponent is 6.

Note that a third order polynomial is not always a trinomial. $2x^3 + 4x$ and $4x^3 + x^2 + 3x + 8$ are both third order polynomials, but neither of them has three terms.

CONCEPTS TO HELP YOU

Several topics covered earlier in the book are essential when working with polynomials: *like terms*, *factors*, and *powers*, in particular, come into play when performing operations with polynomials.

Adding and Subtracting Polynomials

Adding and subtracting polynomials involves the same concepts we reviewed in chapter 2, in connection with simplifying algebraic expressions. The key to adding and subtracting polynomials is to identify like terms. Only like terms can be combined by addition and subtraction, you should recall.

Two terms are like as long as neither contains a variable that the other doesn't have. But now that we are dealing with powers, we have to be more specific. In order for two terms to be like, any variable found in one term must be raised to the same power in both.

$4x^3$ and $10x^3$	Like terms
$9a^2b$ and $-5a^2b$	Like terms
$6y^2$ and $6y^3$	Not like terms

Adding and subtracting like monomials works just as it does for other algebraic terms; you combine the numbers appearing before the variables (and if no number appears, you treat it as though there is a 1):

$$3x^4 + 5x^4 = 8x^4$$
$$7a^3b^2 - 11a^3b^2 = -4a^3b^2$$
$$6a^3b^2 + a^3b^2 = 7a^3b^2$$

Multiplying Polynomials

The key to multiplying polynomials is the distributive property, which we covered in chapter 2. By applying this property, we can pair each term in the first polynomial with each term in the second, and find the product of each pair. Again, the distributive property holds that $(x + y) \bullet z = x \bullet z + y \bullet z$.

Let's take the polynomials $a + b$ and $c + d + e$. Following the distributive property:

$$(a + b) \bullet (c + d + e) = a \bullet (c + d + e) + b \bullet (c + d + e)$$

To simplify this expression, we have to use the distributive property two more times:

$$a \bullet (c + d + e) = ac + ad + ae$$
$$b \bullet (c + d + e) = bc + bd + be$$

So the product is:

$$ac + ad + ae + bc + bd + be$$

This product cannot be further simplified. Unless further simplification is possible, the number of terms in the product of two polynomials is the number of terms in the first polynomial multiplied by the number of terms in the second. There are two terms in $a + b$, three terms in $c + d + e$, and six terms in the product of the polynomial.

If you get a product of a binomial and a trinomial with no similar terms in the product, that has fewer than six terms before you simplify, you must have forgotten to multiply each pair of terms.

The FOIL Method

In Algebra I, multiplication of binomials is very common. In fact, there's a shortcut technique for applying the distributive property. FOIL can help you to pair up terms for multiplication systematically. FOIL stands for the following:

- **First**
- **Outer**
- **Inner**
- **Last**

Take the binomials $x + 2$ and $x + 3$. When we include them in an operation, we can pair them up using the FOIL method.

The FOIL method matches each pair of terms that must be multiplied. The product of $x + 2$ and $x + 3$ is the sum of the following:

- **First:** $x \cdot x = x^2$
- **Outer:** $x \cdot 3 = 3x$
- **Inner:** $2 \cdot x = 2x$
- **Last:** $2 \cdot 3 = 6$

So we get:

$$(x + 2) \cdot (x + 3) = x^2 + 3x + 2x + 6 = x^2 + 5x + 6$$

Dividing Polynomials

Dividing a polynomial by a monomial is like simplifying a fraction. You divide each term in the polynomial by the monomial, following the principles for division laid out in chapters 2 and 4.

You can also divide a polynomial by a binomial, trinomial, or an even longer expression. We'll address this in chapter 12.

Factoring Polynomials

Since many polynomials are the products of other polynomials, you may be asked to identify a polynomial's factors. We'll focus here on monomial and binomial factors.

Monomial Factors

A monomial is a factor of a polynomial if every term in the polynomial can be divided by the monomial with the result being another monomial.

x^2 is a factor of $3x^4 - 2x^3$

[Since $3x^4 \div x^2 = 3x^2$ and $2x^3 \div x^2 = 2x$]

So $(3x^4 - 2x^3) \div x^2 = 3x^2 - 2x$

Binomial Factors

We are especially interested in binomial factors of third–order polynomials. They are very important in quadratic equation-solving, as we'll see in the next chapter.

Let's take the binomials $x + 4$ and $x - 3$, and use the FOIL method to get their product. Using the FOIL method, we'll identify four pairs of terms in the expression $(x + 4) \bullet (x - 3)$:

Then we'll find the product of each pair of terms:

First: $x \bullet x = x^2$

Outer: $x \bullet -3 = -3x$

Inner: $4 \bullet x = 4x$

Last: $4 \bullet 3 = -12$

So:

$(x + 4) \bullet (x - 3) = x^2 - 3x + 4x - 12 = x^2 + x - 12.$

In general, the product of two binomials $x + j$ and $x + k$ is $x^2 + (j+k)x + jk$.

In the case of $(x + 4) \bullet (x - 3)$, $j = 4$ and $k = -3$, and the product of $x + j$ and $x + k$ is $x^2 + (4 - 3)x + (4 \bullet - 3)$.

The sum of j and k is 1, and the product of j and k is -12. Those numbers are the coefficients of the second and third terms of the polynomial $x^2 + x - 12$.

With this in mind, you can often factor third-order trinomials into binomials by working backward. If you find two numbers whose sum equals the second coefficient in a trinomial, and whose product is the third coefficient, then you have identified the binomial factors.

STEPS YOU NEED TO REMEMBER

Here we will go through the steps for combining polynomials. Addition, subtraction, and multiplication are covered here. The division of polynomials is examined in chapter 12, where we will go over algebraic fractions.

Adding and Subtracting

When subtracting a polynomial, first remember that you must distribute the negative sign, as explained in chapter 2.

$$x^2 - (2x^2 + 4x) = x^2 - 2x^2 - 4x$$
$$x^2 - (2x^2 + 4x) \neq x^2 - 2x^2 + 4x$$

When adding and subtracting polynomials and you need to combine like terms, you can organize monomial terms using the associative and commutative properties. $(4a + 5b) + (3a + 7b)$, for instance, can be rewritten and simplified by the following steps:

$$(4a + 5b) + (3a + 7b) =$$
$$(4a + 5b) + 3a + 7b =$$
$$(4a + 3a) + 5b + 7b =$$
$$7a + 5b + 7b =$$
$$7a + 12b$$

When combining polynomials as short as these, you can try to find like terms in your head. With longer polynomials, however, using the associative and commutative properties is quite helpful. They will help you reorganize the polynomials in small steps, ensuring you don't overlook or mismatch any terms.

Multiplying Polynomials

Multiplication of polynomials doesn't involve just combining like terms; in multiplication, *everything* gets combined. This means that every term in your first polynomial gets multiplied by each term in the second. After that, you add up the results and simplify when possible. That is how you use the distributive property in connection with polynomials. FOIL provides a way of mechanically applying the distributive property in one kind of case: the multiplication of binomials.

STEP–BY–STEP ILLUSTRATION OF THE 5 MOST COMMON QUESTION TYPES

Now that we have reviewed the main concepts and steps in connection with polynomials, we can walk through some important question types.

Question 1: Evaluating Polynomials

If $x = -7$, what is the value of $x^2 - 8x + 32$?

(A) -73

(B) 9

(C) 25

(D) 39

(E) 137

This question requires that you evaluate an algebraic expression, but this time, with powers. Evaluating polynomials commonly involves evaluating monomials with exponents. Here, you need to evaluate each monomial term with a variable, and then simplify the expression.

$x^2 = -7 \bullet -7 = 49$

$8x = -56$

$x^2 - 8x + 32 =$

$49 - (-56) + 32 =$

$49 + 56 + 32 =$

$105 + 32 =$

137

So (E) is the correct answer. One might get to a value of 39, (D), by getting -49 instead of 49 for the value of x^2. (The square of a negative number is positive.) One might get -73, (A), by subtracting 56 instead of -56.

Question 2: Adding Polynomials

The sum of $3x^5 + 7x^3 + 8x^2$ and $5x^5 + x^4 + 11x^3$ is

(A) $8x^5 + 18x^3$

(B) $8x^5 + 8x^3 + 19x^2$

(C) $8x^5 + x^4 + 7x^3 + 19x^2$

(D) $8x^5 + x^4 + 18x^3 + 8x^2$

(E) $8x^{10} + 8x^7 + 19x^5$

We're combining two trinomials here, so we need to combine all pairs of like terms. Two terms are like only if they have the same variable raised to the same power. So $3x^5$ and $5x^5$, and $7x^3$ and $11x^3$ are like terms, while $8x^2$ and x^4 are not.

$(3x^5 + 7x^3 + 8x^2) + (5x^5 + x^4 + 11x^3)$ — Reorganize the terms and parentheses so that like ones are next to each other

$(3x^5 + 5x^5) + x^4 + (7x^3 + 11x^3) + 8x^2$

$3x^5 + 5x^5 = 8x^5$ and $7x^3 + 11x^3 = 18x^3$

$8x^5 + x^4 + 18x^3 + 8x^2$

(D) is the correct answer. Choice (A), $8x^5 + 18x^3$, includes only the monomials resulting from combining like terms. Even the monomials that don't get combined are part of the sum. (B) results from combining the second terms in each trinomial, and then the third terms, even though they're both unlike pairs. (C) could be the result an effort to combine like terms, but it adds $11x^3$ instead of $11x^3$.

Question 3: Multiplying Polynomials

What is the product of $y^2 + 2y + 5$ and $y - 3$?

(A) $y^2 + y - 15$

(B) $y^3 + y - 15$

(C) $y^3 - y^2 - y - 15$

(D) $y^3 - y^2 + 5y - 15$

(E) $y^3 + y^2 + y - 15$

This question is challenging. Getting the product of these polynomials is a matter of using the distributive property. You must find the product of every pair of terms consisting of one term from $y^2 + 2y + 5$ and one term from $y - 3$. This is because, by the distributive property,

$$(y^2 + 2y + 5) \bullet (y - 3) = (y^2 + 2y + 5) \bullet y + (y^2 + 2y + 5) \bullet -3$$

So there are six products to calculate:

$y^2 \bullet y = y^3$

$2y \bullet y = 2y^2$

$5 \bullet y = 5y$

$y^2 \bullet -3 = -3y^2$

$2y \bullet -3 = -6y$

$5 \bullet -3 = -15$

Next, we need the sum of these six products. We can arrange them so as to combine like terms effectively.

$y^3 + 2y^2 + 5y + -3y^2 + -6y + -15 =$

$y^3 + 2y^2 - 3y^2 + 5y - 6y - 15 =$

$y^3 - y^2 + y - 15$

(C) is the correct answer. Had you not added exponents when multiplying terms with variables, you might have gotten (A), $y^2 + y - 15$. The product of y^2 and y is taken to be y^2 instead of y^3, and the product of $2y$ and y is taken to be $2y$ instead of $2y^2$. (D) fails simply to include the product of $2y$ and -3 in the sum.

Question 4: Dividing Polynomials

What is $6x^8 + 4x^6$ divided by $2x^2$?

(A) $3x^4 + 2x^3$

(B) $5x^4$

(C) $3x^6 + 2x^4$

(D) $7x^6$

(E) $5x^7$

To divide a binomial by a monomial, divide each term of the binomial by that monomial; the distributive property works for division as well. So $6x^8 + 4x^6 \div 2x^2 = (6x^8 \div 2x^2) + (4x^6 \div 2x^2)$. We need to carry out each division operation, then, and take the sum of the results. Remember that dividing powers involves *subtracting* exponents; dividing the exponents is a common mistake. So let's carry out our division:

$$6x^8 \div 2x^2 = 3x^6$$
$$4x^6 \div 2x^2 = 2x^4$$

$$(6x^8 \div 2x^2) + (4x^6 \div 2x^2) = 3x^6 + 2x^4$$

Since $3x^6$ and $2x^4$ are not like terms, they cannot be combined by addition. The expression is fully simplified, and **(C) is correct**. Choice (A) results from dividing the exponents, as we have cautioned against.

Question 5: Factoring Polynomials

What are the binomial factors of $x^2 + 9x + 20$?

(A) $x + 5$ and $x + 4$

(B) $x + 7$ and $x + 2$

(C) $x + 10$ and $x + 2$

(D) $x + 16$ and $x + 4$

(E) $x + 20$ and $x + 1$

A trinomial such as this is the product of two binomials: $(x + j)$ and $(x + k)$. Answering this question, then, is a matter of figuring out what must be the values of j and k. To multiply one binomial by another, we multiply each term in the first binomial by each term in the second binomial, and add the products.

$$(x + j) \bullet (x + k) =$$
$$(x \bullet x) + (j \bullet x) + (x \bullet k) + (j \bullet k) =$$
$$x^2 + jx + kx + jk$$

Now use the distributive property to combine jx and kx

$$jx + kx = (j + k)x$$
$$x^2 + jx + kx + jk = x^2 + (j + k)x + jk$$

If $x^2 + 9x + 20 = x^2 + (j + k)x + jk$, then we know that j and k have a sum of 9 and a product of 20. At this point, can use trial and error to get to the values of j and k.

Start by listing the factors of 20. What pairs of numbers have 20 as a product? Do any of those pairs have a sum of 9?

In fact, the factors 4 and 5 have 20 as their product, and 9 as their sum. So 4 and 5 are the values of j and k (though we could also say they're the values of k and j, respectively).

So $x + 5$ and $x + 4$ are the binomial factors of $x^2 + 9x + 20$, and **(A) is correct.** $x + 7$ and $x + 2$, the factors in (A), have a sum of 9 as well, though they don't have a product of 20. (C) and (E) have the opposite problem, meeting the product of 20 condition and not that of the sum.

Since the factoring of polynomials plays an important role in solving quadratic equations, we'll revisit this in the next chapter.

CHAPTER QUIZ

1. If $b = -9$, $5b - 2b^2 =$
   (A) -369
   (B) -279
   (C) -207
   (D) -162
   (E) -137

2. If $x = -6$ and $y = 11$, then
   $x^2y + xy^2 + 2xy =$

3. $(6x^3 + 7x^2) + (4x^2 + 8x^3) =$
   (A) $10x^3 + 15x^2$
   (B) $13x^3 + 12x^2$
   (C) $14x^3 + 11x^2$
   (D) $25x^5$
   (E) $25x^{10}$

4. $(4y^3 + 9y^2 - 11y) - (2y^3 - 4y^2 - 3y) =$
   (A) $2y^3 + 5y^2 - 8y$
   (B) $2y^3 + 5y^2 - 14y$
   (C) $2y^3 + 5y^2 + 8y$
   (D) $2y^3 + 13y^2 - 8y$
   (E) $2y^3 + 13y^2 - 14y$

5. What is the product of $x + 5$ and $x - 6$?
   (A) $x^2 - 30x - 11$
   (B) $x^2 - 30x - 1$
   (C) $x^2 - 11x - 30$
   (D) $x^2 - 11x + 30$
   (E) $x^2 - x - 30$

6. What is the product of $3xy - 5x$ and $4xy + 2y - 6$?
   (A) $12x^2y^2 - 14xy^2 - 10xy + 30x$
   (B) $12x^2y^2 + 6xy^2 - 20x^2y$
       $- 10xy + 30x$
   (C) $12x^2y^2 + 6xy^2 - 20x^2y$
       $- 16xy + 30x$
   (D) $18x^2y^2 - 20x^2y - 10xy + 30x$
   (E) $18x^2y^2 - 14x^2y - 6xy + 30x$

7. a and $a - 8$ are both factors of which polynomial?
   (A) $-8a^2$
   (B) $-7a$
   (C) $a^2 - 8$
   (D) $a^2 - 8a$
   (E) $a^2 - 8a - 8$

8. A factor of $y^2 - 17y + 60$ is
   (A) $y + 12$
   (B) $y - 12$
   (C) $y + 5$
   (D) $y - 4$
   (E) $y + 15$

ANSWER EXPLANATIONS

1. C

Use the value of b to get the value of each monomial, and then subtract to get the value of the binomial.

$$5b = 5(-9) = -45$$
$$2b^2 = 2(-9^2) = 2(81) = 162$$
$$5b - 2b^2 = -45 - 162 = -207$$

Choice (A), -369, is actually the value of $5b - (2b)^2$. In the absence of parentheses, the order of operations tells you to evaluate powers before multiplying. (E), 137, results from subtracting 162 from 45 instead of -45.

2. −396

Use the values of the variables to evaluate each monomial. Then add the values to get the value of the trinomial.

$$x^2y = 6^2 \bullet 11 = 36 \bullet 11 = 396$$
$$xy^2 = -6 \bullet 11^2 = -6 \bullet 121 = -726$$
$$2xy = 2 \bullet -6 \bullet 11 = -132$$
$$x^2y + xy^2 + 2xy = 396 + -726 + -132 = -462$$

Be careful when you multiply negative numbers, as that can be tricky. A value of 1056 would be the result of taking $-6 \bullet 11^2$ to equal 726 instead of -726.

3. C

Use the associative and commutative properties to group like terms:

$$(6x^3 + 7x^2) + (4x^2 + 8x^3) =$$
$$(6x^3 + 7x^2) + 8x^3 + 4x^2 =$$
$$(6x^3 + 8x^3) + 7x^2 + 4x^2 =$$
$$14x^3 + 7x^2 + 4x^2 =$$
$$14x^3 + 11x^2$$

Choice (A), $10x^3 + 15x^2$, is the sum of $6x^3 + 7x^2$ and $4x^3 + 8x^2$. Be careful of polynomials where the terms are not always ordered in the same way.

4. D

When subtracting polynomials, remember to distribute the minus sign:

$$(4y^3 + 9y^2 - 11y) - (2y^3 - 4y^2 - 3y) =$$
$$(4y^3 + 9y^2 - 11y) - 2y^3 - (-4y^2) - (-3y) =$$
$$(4y^3 + 9y^2 - 11y) - 2y^3 + 4y^2 + 3y$$

Now we're in a better position to combine terms. Use the associative and commutative properties to group like terms, and simplify:

$$(4y^3 - 2y^3) + (9y^2 + 4y^2) - 11y + 3y =$$
$$2y^3 + 13y^2 - 8y$$

Choice (A) would result from subtracting $4y^2$ instead of adding, and (E) would result from subtracting $3y$ instead of adding.

5. E

Use the FOIL method find the product of these binomials.

First: $x \bullet x = x^2$

Outer: $x \bullet -6 = -6x$

Inner: $5 \bullet x = 5x$

Last: $5 \bullet -6 = -30$

$$x^2 - 6x + 5x - 30 = x^2 - x - 30$$

Choice (C), $x^2 - 11x - 30$, is the result of subtracting $5x$ instead of adding.

6. C

Use the distributive property to set up the multiplication:

$$(3xy - 5x) \bullet (4xy + 2y - 6) = 3xy \bullet (4xy + 2y - 6) - 5x \bullet (4xy + 2y - 6)$$
$$3xy \bullet (4xy + 2y - 6) = 12x^2y^2 + 6xy^2 - 6xy$$
$$5x \bullet (4xy + 2y - 6) = 20x^2y + 10xy - 30x$$

$$(3xy - 5x) \bullet (4xy + 2y - 6) =$$
$$(12x^2y^2 + 6xy^2 - 6xy) - (20x^2y + 10xy - 30x) =$$
$$12x^2y^2 + 6xy^2 - 20x^2y - 6xy - 10xy + 30x =$$
$$12x^2y^2 + 6xy^2 - 20x^2y - 16xy + 30x$$

So (C) is correct. (A) results from subtracting $20xy^2$ instead of $20x^2y$, and not subtracting $6xy$ at all.

7. D

a and $a - 8$ are both factors of many polynomials. One polynomial in particular that's easy to identify as one with both of these factors is the product of a and $a - 8$:

$$a \bullet (a - 8) = a \bullet a + a \bullet -8 = a^2 - 8a$$

You can also factor a out of each expression in the answer choices by dividing each by a. If the result is polynomial, then a is a factor. You would check to see if the result has $a - 8$ as a factor. Since $(a^2 - 8a) \div a = a - 8$, we see that both a and $a - 8$ are factors.

8. B

Since the polynomial $y^2 - (j + k)y + j$ is the product of its factors $y + j$ and $y + k$, we can identify those factors by finding two numbers that have a product of 60 and a sum of -17. If two numbers have a negative sum, at least one of them must be negative. If they have a positive product, they must both be negative (for the product of a negative number and a positive one is negative). Two negative numbers with a product of 60 are -12 and -5, and the sum of those is -17. So $y - 12$ and $y - 5$ are both factors of 60.

(A), $y + 12$, and Choice (C), $y + 5$, are both tempting because 12 and 5 have a sum of 17 and a product of 60. However, we need a sum of -17.

Quadratic Equations

WHAT ARE QUADRATIC EQUATIONS?

"Quadratic" is a special term for a second–order polynomial. A quadratic equation is an equation of the form $ax^2 + bx + c = d$. Actual quadratic equations usually appear with numbers in place of the letters $a, b, c,$ and d.

Quadratic equations get special attention in algebra because many of them cannot be solved with the methods we have covered so far. Many of them feature trinomials, though many do not.

Any of the letters in a quadratic equation can have a value of 0. If $a = 2$, $b = 0$, $c = 4$, and $d = 5$, for instance, we have the quadratic equation $2x^2 + 4 = 5$ (since $b = 0$, $bx = 0$). As long as the number a represents is not 0, we have to deal with the equation as a quadratic equation.

In this chapter, we will look at two methods for solving quadratic equations.

CONCEPTS TO HELP YOU

Many quadratic equations have two solutions. The reason they yield two solutions is because they have a form that doesn't lend itself to the steps for solving linear equations. Factoring is a method that plays a major role in quadratic equation-solving. Even if we don't use factoring, however, and opt for the quadratic formula explained below, the roots of a quadratic equation are closely tied to the factors of the quadratic.

Quadratic solutions are commonly referred to as roots.

Standard Form

The "standard form" of a quadratic equation is $ax^2 + bx + c = 0$. That's the original form shown above, where $d = 0$. An equation can be written as a standard form quadratic equation as long as it's possible to get a second–order polynomial on one side and 0 on the other. Just as you worked with linear equations to get them into slope–intercept form in chapter 7, you can work with quadratic equations to get them into standard form. This is important, because the methods for solving quadratic equations we will discuss work with equations in that standard form.

Solving by Factoring

Many quadratics can be factored into a pair of polynomials. These factors can be binomials or monomials. That quadratics can be factored in this way is very important in this discussion.

Take the quadratic equation $ax^2 + bx + c = 0$. Suppose $ax^2 + bx + c$ is the product of $x - j$ and $x - k$. Since the product of two expressions is 0 if one of the expressions has a value of 0, then:

$$(x - j) \bullet (x - k) = 0 \quad \text{If } x - j = 0 \text{ or } x - k = 0$$

Because you get 0 by subtracting any number from itself:

$$x - j = 0 \quad \text{If } x = j$$
$$x - k = 0 \quad \text{If } x = k$$

So when you factor the polynomial in $ax^2 + bx + c = 0$ into $x - j$ and $x - k$, j and k are the solutions to the quadratic equation.

In cases where a is a number other than 1, more work is involved. We'll explain the extra steps shortly.

Binomial Squares

While many quadratic equations have two solutions, many have just one. Since a quadratic equation has two roots (j and k) the equation has one solution if $j = k$. So a quadratic equation with one solution could be rewritten as $(x - j) \bullet (x - j) = 0$, or $(x - j)^2 = 0$. In fact, every quadratic equation with a single solution involves a quadratic that is the square of a binomial. That is why we say that some quadratic equation involve binomial squares.

Some examples of binomial squares:

$(x - 5)^2 = x^2 - 10x + 25$

$(x - 3)^2 = x^2 - 6x + 9$

$(x + 2)^2 = x^2 + 4x + 4$

$(x + 6)^2 = x^2 + 12x + 36$

There is one pattern we can recognize here that will help us to identify quadratic equations with binomial squares in the future. You have a binomial square if $\left(\dfrac{b}{2}\right)^2 = c$.

The Quadratic Formula

Some quadratic equations are simply too difficult to solve by factoring, such as when they have roots that can only be expressed as radicals. If you know that the solution will involves simplified radicals, or you have trouble identifying the binomial factors of a quadratic, you can use the *Quadratic Formula*:

$$x = \frac{-b \pm \sqrt{b^2 - 4ac}}{2a}$$

The Quadratic Formula provides the roots of a quadratic equation on the basis of the values of the coefficients you plug in. The letters a, b, c are the coefficients. Let's take the equation $x^2 + 15x + 4 = 0$.

$x^2 + 15x + 4 = 0$ So $a = 1$, $b = 15$, and $c = 4$. Simply evaluate for these values.

$$\frac{-b \pm \sqrt{b^2 - 4ac}}{2a}$$

If the quadratic has only two terms, the coefficient of the missing term is 0. In $x^2 - 81 = 0$, for instance, $b = 0$. In $x^2 + 5x = 0$, $c = 0$.

You can evaluate $\dfrac{-b \pm \sqrt{b^2 - 4ac}}{2a}$ to get two solutions. That's because the "$\pm$" sign in the formula stands for both addition and subtraction. The expression $4 \pm \sqrt{2}$, for example, means $4 + \sqrt{2}$ and $4 - \sqrt{2}$. So an expression with the "$\pm$" sign is actually two expressions rolled into one.

SOME QUADRATIC EQUATIONS HAVE NO SOLUTIONS

You cannot evaluate radicals containing negative numbers. If you get a negative value for $b^2 - 4ac$ in the course of solving a quadratic equation, a mistake was likely made. An Algebra I question won't ask you to solve a quadratic equation that has no real number solutions.

What Do the Solutions Mean?

While the graph of a linear equation on the coordinate plane is a straight line, the graph of a quadratic equation is a *parabola*—a curved, U-shaped figure set right-side up or upside down. We won't say much about graphing quadratic equations here, but it's important to know that the solutions of quadratic equations are their x–intercepts when the quadratic = 0, or $y = 0$. The parabola below is the graph of a quadratic equation.

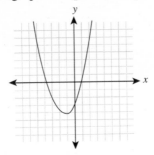

This parabola crosses the x–axis at $(-3, 0)$ and $(1, 0)$. So -3 and 1 are the intercepts or roots of the equation. Since the factors of the quadratic are $x + 3$ and $x - 1$, the quadratic equation is $x^2 + 2x - 3 = 0$.

Since the solutions are x–intercepts or roots, you might be asked to provide the solution to a quadratic equation in those terms.

STEPS YOU NEED TO REMEMBER

Once you have a quadratic equation in its standard form, you might use factoring or the quadratic formula to find the solution(s). Both are extremely useful, and you should be prepared to use either, depending on the equation you encounter.

Standard Form

Remember that a quadratic equation must be in the standard form $ax^2 + bx + c = 0$ before you use the methods we have reviewed. If the equation you are working with is not in standard form, you must first perform the operations necessary to isolate 0 on one side.

Factoring

When the coefficient of the first term in a quadratic equation is 1, you can try to factor the polynomial into two binomials: $x - j$ and $x - k$. If you are successful, then you can conclude that j and k are the solutions. To find the values of j and k, note that:

$$x^2 + bx + c = x^2 + -(j + k)x + jk$$
$$b = -(j + k) \text{ or } -b = j + k$$
$$c = jk$$

To identify the values of j and k, then, you must find the two numbers that have a sum of $-b$ and a product of c.

When the coefficient of the first term is some number *other than 1*, you have two options: (1) You can try to factor into binomials right away, or (2) You can try to factor that coefficient out of the entire quadratic. Let's take an example:

$2x^2 + 8x + 6 = 0$ Factor 2 out of the polynomial, as per the distributive property

$$2(x^2 + 4x + 3) = 0 \quad \text{Divide both sides of the equation by 2. (0 divided by 2 is 0)}$$

$$x^2 + 4x + 3 = 0 \quad \text{Now we have a quadratic that is easier to factor}$$

It won't always be easier to factor a quadratic after factoring out a number this way. Doing so could leave terms with coefficients that are fractions. Sometimes it's easier to factor the quadratic into binomials in the first step. In that case, we'll get at least one binomial with a variable that has a coefficient other than one. In the case of $2x^2 + 8x + 6$, we'd have factors of $2x + 2$ and $x + 3$. The j and k here had to be numbers with a product of 6 that also satisfied the equation $j + 2k = 6$, since k had to be multiplied by $2x$.

To get the solutions using j and k, you must divide those numbers by the variable coefficients in their respective binomials:

$$2x^2 + 8x + 6 = (2x + 2) \bullet (x + 3) = 0$$
$$2x^2 + 8x + 6 = 0 \text{ if } 2x + 2 = 0$$

So the quadratic is 0 if $x = -1$. Since $j = -2$, we had to divide that number by 2, the coefficient of $2x$ to get that solution. Since the coefficient of the variable in the binomial with k is 1, the value of k is the other solution.

Let's introduce two more coefficients: d and e. They'll go in front of the variables of the binomial factors. In general,

$$(dx - j) \bullet (ex - k) = ax^2 + bx + c$$
$$de = a$$
$$-dk - ej = b$$
$$jk = c$$

You may encounter equations with quadratics with factors that have large coefficients. When that happens, getting the correct values of d, e, j and k may be matter of trial and error. Take different pairs of numbers that have a product of a and different pairs of numbers that have a product of d and e. Try to find the combination that satisfies that equation $dk + ej = -b$

The Quadratic Formula

If your calculations become too cumbersome, you might use the Quadratic Formula instead. You can use it with any quadratic equation, in fact, but it is especially handy when it looks like the solutions will involve radicals or fractions. If you use the formula, you need only follow the steps for evaluating radicals and other algebraic expressions.

One of the most important steps—one that can lead to serious errors if done incorrectly—is plugging the right values for a, b, and c into the right spots in the formula.

STEP–BY–STEP ILLUSTRATION OF THE 5 MOST COMMON QUESTION TYPES

Now we learn how to apply factoring and the quadratic formula to real questions. Hopefully, you will get a sense of how one method is better suited for certain equations than the other.

Question 1: Solving by Factoring

If $x^2 - 16x + 64 = 0$, then $x =$

(A) −18 or 2

(B) −8 or 8

(C) 4 or 12

(D) 4

(E) 8

A quadratic equation $ax^2 + bx + c = 0$ has solutions j and k if $(x - j) \bullet (x - k) = ax^2 + bx + c$. So we can find the values of j and k by factoring the quadratic into two binomials.

$(x - j) \bullet (x - k) = x^2 - (j + k)x + jk$

If $x^2 - (j + k)x + jk = x^2 - 16x + 64$, then $j + k = 16$ and $jk = 64$

What pair of numbers has this product? Well, since the sum and the product are positive, j and k must both be positive. So let's list the pair of positive numbers that have a product of 64, along with their sums:

1, 64; $1 + 64 = 65$

2, 32; $2 + 32 = 34$

4, 16; $4 + 16 = 20$

8, 8; $8 + 8 = 16$

This last pair, 8 and 8, meets our requirements; the numbers have a sum of 16 and a product of 64. So $j = 8$ and $k = 8$. We can check this by finding the product of $(x - 8)$ and $(x - 8)$. Using FOIL, we get:

$(x - 8) \bullet (x - 8) = (x \bullet x) + (x \bullet -8) + (x \bullet -8) + (-8 \bullet -8) =$

$x^2 - 8x - 8x + 64 = x^2 - 16x + 64$

Since $j = k$, the two solutions to $x^2 - 16x + 64 = 0$ are one and the same number. There is only one solution, 8, and so **(D) is correct**. Choice (A), -18 or 2, might be tempting because those numbers have a sum of -16. In fact, those numbers are solutions to $x^2 + 16x - 36 = 0$. (B) is also tempting because it includes the number 8, and it presents two solutions; that is certainly more typical for quadratic equations. The quadratic equation that actually has both -8 and 8 as solutions, however, is $x^2 + 64$.

Question 2: Solving Equations Not in Standard Form

What are the solutions of $2x^2 = 56 - 6x$?

(A) -28 and 2

(B) -14 and 8

(C) -8 and 7

(D) -7 and 4

(E) -2 and 14

This quadratic equation is not in the standard form $ax^2 + bx + c = 0$. Before we use factoring to solve, we must get it into that form. This means getting both terms off the right of equation, leaving nothing by 0.

$2x^2 = 56 - 6x$ Add $6x$ to both sides

$2x^2 + 6x = 56$ Subtract 56 from both sides

$2x^2 + 6x - 56 = 0$

We could try to factor at this point, though it might be tricky since the first coefficient is 2. When the factors are $2x - j$ and $x - k$ instead of $x - j$ and $x - k$, it is harder to figure out the values of j and k.

Fortunately, there is something else we can do before factoring. Notice that each of the coefficients is even. That means 2 is a factor of each term. With the help of the distributive property, then, we can factor 2 out of the quadratic altogether:

$2x^2 + 6x - 56 =$

$2(x^2) + 2(3x) - 2(28) =$

$2(x^2 + 3x - 28)$

$2(x^2 + 3x - 28) = 0$ Divide both sides by 2

$x^2 + 3x - 28 = 0$

Factoring this quadratic should be much more straightforward:

$x^2 + 3x - 28 = (x - j) \bullet (x - k)$

$-(j + k) = 3$

$j + k = -3$

$jk = -28$

Choice (D) is the answer; -7 and 4 have a product of -28 and a sum of -3. Choice (A) might be tempting because -28 and 2 have a product of 56. The same goes for (C). With (B), the two numbers have a sum of -6.

Question 3: Advanced Factoring

What are the solutions of $2x^2 + 5x - 12 = 0$?

(A) $-\frac{3}{2}$ and -4

(B) $-\frac{3}{2}$ and 4

(C) $\frac{3}{2}$ and -4

(D) 6 and -4

(E) 6 and 4

Since the value of a in this equation is not 1, finding the solutions will take factors are
$(2x - j)(x - k)$.

So we know that $j + 2k = -5$ and $jk = -12$. One pair of factors of -12 is -4 and 3. If you double -4 and add 3, you get -5. So $j = 3$ and $k = -4$.

Since $(2x - j)(x - k) = 0$ if $2x - j = 0$, one solution is $x = \frac{j}{2} = \frac{3}{2}$.

So $x = \frac{3}{2}$ or $x = -4$, and **(C) is correct.** Choice (B) might be tempting because its numbers are based on the values of $-j$ and $-k$. Choice (D), 6 and -4, would result from multiplying instead of dividing j by 2.

Question 4: Using the Quadratic Formula

If $x^2 + 8 - 8 = 0$, $x =$

(A) $-4 \pm 4\sqrt{2}$

(B) $-4 \pm 2\sqrt{6}$

(C) $-4 \pm 6\sqrt{3}$

(D) $4 \pm 4\sqrt{2}$

(E) $4 \pm 6\sqrt{3}$

When you see a quadratic equation question where all the answer choices involve radicals, it's safe to assume that solving by factoring would be difficult. Moreover, if you have trouble identifying a pair of numbers that have a sum of –8 (as they must, since $-j + -k = 8$) and a product of –8, then factoring would take a long time. So this equation is better suited for the quadratic formula:

$$x = \frac{-b \pm \sqrt{b^2 - 4ac}}{2a}$$

$ax^2 + bx + c = x^2 + 8 - 8 \qquad a = 1, b = 8, c = -8$. Plug these values into the formula

$$x = \frac{-8 \pm \sqrt{8^2 - 4(1)(-8)}}{2(1)} =$$

$$\frac{-8 \pm \sqrt{64 - (-32)}}{2} =$$

$$\frac{-8 \pm \sqrt{96}}{2} =$$

$$-4 \pm \sqrt{\frac{96}{2}} =$$

$$-4 \pm \sqrt{\frac{(16)(6)}{2}} =$$

$$-4 \pm \frac{4\sqrt{6}}{2} =$$

$$-4 \pm 2\sqrt{6}$$

This last value is the fully simplified solution, and **(B) is the correct answer**. (A) would result from subtracting 32 from 64 instead of adding, thereby getting $-4 \pm \sqrt{32}$ instead of $-4 \pm \sqrt{96}$. (C) would result from taking 96 to be the product of 36 and 3 (that product is actually 108).

Question 5: Using the Roots to Derive Equations

Which equation has a graph with x–intercepts –3 and 6?

(A) $x^2 - 18x - 3 = 0$

(B) $x^2 - 9x - 18 = 0$

(C) $x^2 - 3x - 18 = 0$

(D) $x^2 + 3x - 18 = 0$

(E) $x^2 + 6x - 9 = 0$

The roots of a quadratic equation are the x–intercepts of its graph—the points where it crosses the x–axis. Since parabolas are curved, many of them intercept the x–axis at two points. The quadratic equation can be expressed in terms of the product of two binomial factors $x - j$ and $x - k$.

Since the solutions of the equation are -3 and 6, let's say that $j = -3$ and $k = 6$.

$(x - j) \bullet (x - k) =$

$(x + 3) \bullet (x - 6) =$

$x^2 - 6x + 3x - 18 =$

$x^2 - 3x - 18$

So the equation with solutions -3 and 6 is $x^2 - 3x - 18 = 0$, and **(C) is correct**. Choice (B), $x^2 - 9x - 18 = 0$, is the result of getting $-3x$ instead of $3x$ when carrying out FOIL. Choice (D), $x^2 + 3x - 18 = 0$, is the result of multiplying the binomials $x - 3$ and $x + 6$ instead of $x + 3$ and $x - 6$.

CHAPTER QUIZ

1. Which of the following equations has exactly one solution?

   (A) $x^2 - 64 = 0$

   (B) $x^2 - 14x + 49 = 0$

   (C) $x^2 - 9x + 81 = 0$

   (D) $x^2 + 16x + 48 = 0$

   (E) $x^2 + 34x + 64 = 0$

2. The solution of
   $x^2 + 20x + 100 = 0$ is

3. If $x^2 + 11x - 42 = 0$, then $x =$

   (A) -14 and 3

   (B) -14 and -3

   (C) -13 and 2

   (D) -7 and 6

   (E) -6 and -7

4. If $-x^2 - 17x + 60 = 0$, then $x =$

   (A) -20 and 3

   (B) -12 and -5

   (C) 12 and 5

   (D) 12 and -5

   (E) 20 and -3

5. If $3x^2 + 24x - 144 = 0$, then $x =$
   (A) -36 and -4
   (B) -36 and 4
   (C) -12 and 4
   (D) -12 and 12
   (E) 36 and -4

6. If $2x^2 - 7x + 2 = 0$, what is the value of x?
   (A) $\dfrac{-7 \pm \sqrt{33}}{4}$
   (B) $\dfrac{-7 \pm \sqrt{65}}{4}$
   (C) $\dfrac{7 \pm \sqrt{33}}{4}$
   (D) $\dfrac{7 \pm \sqrt{33}}{2}$
   (E) $\dfrac{7 \pm \sqrt{65}}{2}$

7. If $8x - 3 = 3x^2$, then $x =$
   (A) $\dfrac{-4 \pm \sqrt{7}}{3}$
   (B) $\dfrac{-4 \pm \sqrt{5}}{3}$
   (C) $\dfrac{4 \pm \sqrt{7}}{3}$
   (D) $\dfrac{4 \pm 2\sqrt{5}}{3}$
   (E) $\dfrac{4 \pm 2\sqrt{7}}{3}$

8. Which of the following equations has x–intercepts of $\dfrac{3}{2}$ and -5?
   (A) $2x^2 - 13x + 15 = 0$
   (B) $2x^2 - 7x - 15 = 0$
   (C) $2x^2 - 7x + 15 = 0$
   (D) $2x^2 + 7x - 15 = 0$
   (E) $2x^2 + 13x + 15 = 0$

ANSWER EXPLANATIONS

1. B

If a quadratic equation is the square of a binomial, it has one solution. You have such a square if $\left(\dfrac{b}{2}\right)^2 = c$. The only equation where c is the square of half of b is $x^2 - 14x + 49 = 0$, since 49 is the square of -7, which is half of -14. Choice (A), $x^2 - 64 = 0$, might be tempting because the solutions, 8 and -8, have the same absolute value. As a matter of fact, no quadratic that is missing a middle term is the square of a binomial. (C), $x^2 - 9x + 81 = 0$, might also be tempting, since the second and third terms are perfect squares. In fact, that equation does not have real solutions.

2. −10

Your clue is the singular word *solution*. Only equations with polynomials that are squares of binomial factors have a single solution. You can confirm that you have a binomial square by noting that since $10^2 = 100$, $\left(\frac{b}{2}\right)^2 = c$. So $x^2 + 20x + 100 = (x + 10)^2$, or $(x + 10) \bullet (x + 10)$. Since $(x + 10) \bullet (x + 10) = 0$ if $x = -10$, that is the solution.

The number 10, what you get in the binomial, is a tempting solution. But you need to find the value of j where $x − j$, not $x + j$, equals $x + 10$.

3. A

If j and k are the solutions of $x^2 + 11x − 42 = 0$, then $(x − j) \bullet (x − k) = x^2 + 11x − 42$. So $−(j + k) = 11$, , or $j + k = −11$, and $jk = −42$. The two numbers that have a product of −42 and a sum of −11 are −14 and 3. Choice (C) might be tempting because −13 and 2 have a sum of −11. Since their product is −26, however, they are actually the solutions to $x^2 + 11x − 26 = 0$. (D) is tempting as well because −7 and 6 have a product of −42. Their sum is −1, however, so they are the solutions to If $x^2 + x − 42 = 0$.

4. A

With this quadratic equation, the first coefficient is negative. Factoring it is a bit tricky, but it can be done. An easier alternative is to factor out −1.

$$-x^2 - 17x + 60 = 0$$
$$-1(x^2 + 17x - 60) = 0 \qquad \text{Divide both sides by } -1$$
$$x^2 + 17x - 60 = 0$$
$$x^2 + 17x - 60 = (x + 20) \bullet (x - 3)$$

So the solutions are −20 and 3.

Choice (E), 20 and −3, shows the solutions for a similar equation, $−x^2 + 17x + 60 = 0$.

5. C

You might decide to use the quadratic formula here, but factoring is by no means impossible. Since $a = 3$, the factors of $3x^2 + 24x - 144$ should be $3x - j$ and $x - k$. So what are j and k? Since $(3x - j) \bullet (x - k) = 3x^2 - (j + 3k)x + jk$, $j + 3k = -24$, and $jk = -144$.

Listing all of the pairs of numbers that have a product of -144 and sum of -24, we can pick out -36 and 4 as the pair that satisfies $j + 3k = -24$. If $j = -36$ and $k = 4$, then $j + 3k = -36 + 3(4) = -36 + 12 = -24$. Since $(3x + 36) \bullet (x - 4) = 0$ if $x = -12$ or $x = 4$, those are the solutions.

Choice (B), -36 and 4, is probably the most tempting incorrect answer, since those numbers are the values of j and k we came up. When the value of a is something other than 1, however, division is usually necessary to get to the real solutions.

6. C

You can answer this question with the help of the quadratic formula, where $a = 2, b = -7, c = 2$:

$$x = \frac{-b \pm \sqrt{b^2 - 4ac}}{2a}$$
$$= \frac{-(-7) \pm \sqrt{(-7)^2 - 4(2)(2)}}{2(2)}$$
$$= \frac{7 \pm \sqrt{49 - 16}}{4}$$
$$= \frac{7 \pm \sqrt{33}}{4}$$

Choice (D), $\frac{7 \pm \sqrt{33}}{2}$ is the result of using 2 instead of $2a$ as the denominator in the quadratic formula. *Choice* (E), $\frac{7 \pm \sqrt{65}}{4}$ is the result of adding 16 to the 49 inside the radical sign, instead of subtracting.

7. C

First you'll need to get this in the standard form of a quadratic equation. If $8x - 3 = 3x^2$, then $-3x^2 + 8x - 3 = 0$. It doesn't look like this will be easy to solve by factoring, so use the quadratic formula. Since $a = -3$, $b = 8$, $c = -3$:

$$x = \frac{-b \pm \sqrt{b^2 - 4ac}}{2a}$$

$$x = \frac{-b \pm \sqrt{b^2 - 4ac}}{2a}$$

$$= \frac{-8 \pm \sqrt{8^2 - 4(-3)(-3)}}{2(-3)}$$

$$= \frac{-8 \pm \sqrt{64 - 36}}{-6}$$

$$= \frac{-8 \pm \sqrt{28}}{-6}$$

$$= \frac{-8 \pm \sqrt{(4)(7)}}{-6}$$

$$= \frac{-8 \pm 2\sqrt{7}}{-6}$$

$$= \frac{4 \pm \sqrt{7}}{3}$$

Choice (A), $\frac{-4 \pm \sqrt{7}}{3}$ is the solution of $-3x^2 - 8x - 3 = 0$. Choice (E) would result had you not eliminated the 2 when dividing the numerator and denominator by -2.

8. D

A quadratic equation with solutions j and k factors can be rewritten $(x - j)(x - k) = 0$. So the equation we are looking for can be derived from $(x - \frac{3}{2} + 5) = 0$. Since the equation starts with $2x^2$, we should double one of the factors. If $x - \frac{3}{2}$ 0, then $2x - 3 = 0$. So $(2x - 3)(x + 5) = 0$.

By multiplying those factors, we get $2x^2 + 10x - 3x - 15 = 2x^2 + 7x - 15 = 0$. Choice (A) is the result of multiplying $2x - 3$ and $x - 5$. That equation would have the solutions $\frac{3}{2}$ and 5. Choice (B) is the result of $2x + 3$ and $x - 5$.

Algebraic Fractions

WHAT ARE ALGEBRAIC FRACTIONS?

An *algebraic fraction* is a fraction with variables—in the numerator, the denominator, or both. In this chapter, we'll focus on simplifying such fractions, combining them, and solving equations involving them. Those tasks will require us to apply a great deal of what we learned in earlier chapters.

Algebraic fractions are not polynomials, even though they involve polynomials. Recall from chapter 10 that a polynomial cannot be a fraction with a variable in the denominator. You can think of an algebraic fraction as an expression where polynomials are divided by one another.

CONCEPTS TO HELP YOU

We will have to recall some of the concepts involved in factors, as well as the principles for multiplying and dividing polynomials. Finally, working with equations involving algebraic fractions may also involve quadratic equation-solving.

Undefined Algebraic Fractions

A fraction is undefined when the denominator equals 0. Likewise, an algebraic fraction can be undefined for certain variables of its variables. The fraction $\frac{1}{2-x}$, is undefined when $x = 2$, for instance, because $2 - x = 0$ for that value of x. Note that fractions have a value of 0 if the numerator has that value. A value of 0 and an undefined value are not the same thing.

Dividing Polynomials

The division of polynomials is an important operation in connection with algebraic fractions. We'll focus on division of monomials and the use of factoring to divide polynomials.

The key to dividing monomials is handling powers. We know that powers can be divided if you subtract their exponents. When it comes to algebraic terms, however, powers must have the same variable. So you can divide x^5 by x^4, but not by y^4.

When you have a monomial involving several variables, you may be able to divide them by breaking up the terms into factors, and dividing separately. If we want to simplify the following expression, as a matter of dividing the numerator by the denominator:

$$\frac{x^5 y^4}{x^4}$$

Take one of the factors of $x^5 y^4$, x^5, and divide it by x^4

$$x^5 \div x^4 = x^{5-4} = x^1 = x$$
$$x^5 y^4 \div x^4 = x y^4$$

Sometimes you will divide a monomial by one of a higher power. The result will be an expression with a negative power: $x^3 \div x^5 = x^{3-5} = x^{-2}$. A power with a negative exponent is actually a fraction, with the variable in the denominator raised to the absolute value of the exponent you got by subtraction: $x^{-2} = \frac{1}{x^2}$.

Why do negative exponents work this way? Consider the fraction $\frac{8}{32}$. It can be simplified to $\frac{1}{4}$. It could also be rewritten as $\frac{2^3}{2^5}$, since $8 = 2^3$ and $32 = 2^5$. Since $2^3 \div 2^5 = 2^{3-5} = 2^{-2}$, 2^{-2} must equal $\frac{1}{4}$. That is easier to see if we find that $2^{-2} = \frac{1}{2^2}$.

When it comes to dividing other polynomials such as binomials and trinomials, you can divide as long as one expression is a factor of the other. Since $2x$ is a factor of $2x^2 + 4x$, you can divide $2x^2 + 4x$ by $2x$. Think of it as matter of common factors canceling out:

$$\frac{2x^2 + 4x}{2x} = \frac{2x(x + 2)}{2x}$$

Divide the top and bottom by $2x$ to eliminate this common factor

$$x + 2$$

$$(2x^2 + 4x) \div 2x = x + 2.$$

Simplifying Algebraic Fractions

As explained in chapter 3, many fractions can be simplified by dividing both the numerator and denominator by a common factor. Thus, you can try to find a common factor in the top and bottom of an algebraic fraction. When you identify a common factor, you simply divide each part of the fraction by it, and the simplified fraction is what you have left over.

So simplifying an algebraic fraction is a matter of the numerator and denominator into polynomials, one of which is common to the top and bottom. That factor cancels out, leaving the remaining polynomials as parts of a simplified fraction. We can simplify $\frac{4x^2 - 6}{8x - 12}$ by factoring each binomial, so as to identify a common factor:

$$\frac{4x^2 - 6x}{8x - 12} = \frac{2x(2x - 3)}{4(2x - 3)}$$

Eliminate $2x - 3$ since it is a common factor

$$\frac{2x}{4} = \frac{x}{2}$$

Adding and Subtracting Algebraic Fractions

You know that fractions can be added or subtracted only if the denominators are the same. Once you have common denominators, you can add or subtract the numerators as you would combine other polynomials.

Solving Equations with Algebraic Fractions

Solving an equation with algebraic fractions usually involves multiplication, done for the purpose of getting rid of denominators. Once you multiply the polynomials, you're left with an equation whose solution requires the methods of quadratic equation-solving.

STEPS YOU NEED TO REMEMBER

Undefined Fractions

When it comes to undefined values, you need only be concerned with the denominator of a fraction. If you are asked to determine the value of a variable for which an algebraic fraction is undefined, create an equation with 0 on one side and the denominator on the other. By solving that equation, you are finding the value of the variable that makes the denominator equal 0 and, hence, makes the fraction undefined.

Common Denominators

Common denominators are necessary for adding and subtracting algebraic fractions. Finding common denominators is just a matter of finding a common multiple of the denominators you are given. You can always get a common multiple by getting the product of the polynomial denominators. If one of the denominators happens to be a factor of the other, you can multiply the first to get a denominator that matches the second.

Dividing Monomials

When you divide a variable term like x^3 by a term with a larger exponent, such as x^5, the result is a term with an exponent the absolute value of the difference, which goes *in the denominator* of a fraction.

Factoring

When you factor polynomials for the purpose of simplifying algebraic fractions, you are looking for common factors in the numerator and the denominator. Since many polynomials can be factored in more than one way, you may have to try different ways before you can locate a common factor. You cannot factor just a part of a polynomial; to find a factor of the numerator or denominator, you must factor the entire polynomial.

STEP–BY–STEP ILLUSTRATION OF THE 5 MOST COMMON QUESTION TYPES

Working with algebraic fractions doesn't necessarily involve equation-solving. Some algebraic fraction questions involve simplification. We'll take you through the steps involved in various simplification problems, as well as equation solving.

Question 1: Undefined Algebraic Fractions

The expression $\dfrac{x^2 + 13x + 42}{x^2 - 17x + 60}$ is undefined if $x =$

(A) −7

(B) −5

(C) 6

(D) 12

(E) 15

An algebraic fraction is undefined if the denominator has a value of 0. To answer this question, then, we need to solve the equation $x^2 - 17x + 60 = 0$. As a quadratic equation, this can be expected to have two solutions. Since the question asks for just a single number, we have to select the number that matches one of the solutions we find.

$$x^2 - 17x + 60 \qquad \text{Factors are } (x - j) \text{ and } (x - k)$$

Since the product of these factors is $x^2 - (j + k)x + jk$, we're looking for two numbers that have a sum of 17 and a product of 60. The pairs of numbers that have a product of 60 are:

1, 60

2, 30

3, 20

4, 15

5, 12

6, 10

Only the fifth pair, –5 and –12, has a sum of –17. So $j = 5$ and $k = 12$. Since $(x - j)(x - k) = 0$ if $x = j$ or $x = k$, $x^2 - 17x + 60 = 0$ if $x = 5$ or $x = 12$. So the denominator of the fraction has a value of 0 if it has one of those values, which includes **(D), the correct answer.**

Choice (A), –7, is one of the solutions to the equation $x^2 + 13x + 42 = 0$, which involves the numerator of the fraction. When the variable has that value, the fraction has a value of 0, as opposed to being undefined. (B), –5, is the negative of one the solutions of $x^2 - 17x + 60 = 0$. Under the method we've used, you need to find the negatives of the numbers that have a sum of –17 and a product of 60.

Question 2: Adding Algebraic Fractions

$$\frac{y - 1}{y + 5} + \frac{2y + 1}{y + 2} =$$

(A) $\dfrac{3y^2 + 6y}{y^2 + 7y + 10}$

(B) $\dfrac{3y^2 + 7y + 3}{y^2 + 7y + 10}$

(C) $\dfrac{3y^2 + 9y - 3}{y^2 + 7y + 10}$

(D) $\dfrac{3y^2 + 10y + 3}{y^2 + 7y + 10}$

(E) $\dfrac{3y^2 + 12y + 3}{y^2 + 7y + 10}$

To add these fractions, they must be rewritten with a common denominator. To find one, we can multiply the binomials in the denominators: $y + 5$ and $y + 2$. If we are to multiply each denominator by a binomial, then we must multiply the numerator of that fraction by the same binomial.

So in order to get two fractions with common denominators, we must multiply the top and bottom of $\frac{y-1}{y+5}$ by $y + 2$ and the top and bottom of $\frac{2y+1}{y+2}$ by $y + 5$:

$$\left(\frac{y-1}{y+5} \bullet \frac{y+2}{y+2} \right) + \left(\frac{2y+1}{y+2} \bullet \frac{y+5}{y+5} \right) =$$

$$\frac{y^2 + y - 2}{y^2 + 7y + 10} + \frac{2y^2 + 11y + 5}{y^2 + 7y + 10}$$

Now that we have common denominators, we can add the two fractions. That is simply a matter of adding the numerators: $y^2 + y - 2$ and $2y^2 + 11y + 5$. We can carry out that addition as we added polynomials in chapter 10, by grouping like terms:

$$(y^2 + y - 2) + (2y^2 + 11y + 5) =$$
$$(y^2 + 2y^2) + (y + 11y) + (-2 + 5) =$$
$$3y^2 + 12y + 3$$

So the sum of the fractions is $\frac{3y^2 + 12y + 3}{y^2 + 7y + 10}$, and **(E) is the correct answer.**

Choice (A) is the result of multiplying $2y + 1$ by $y + 2$ instead of $y + 5$. (C) is what you would get by multiplying $y - 1$ by $y + 5$ instead of $y + 2$, and multiplying $2y + 1$ by $y + 2$ instead of $y + 5$.

Question 3: Simplifying Monomial Fractions

$$\frac{10a^2b^8c^5}{2a^5b^6c^6} =$$

(A) $\dfrac{5b^2}{a^3c}$

(B) $\dfrac{5ab^2c}{a^3c}$

(C) $\dfrac{5b^2}{a^7c^{14}}$

(D) $5a^3b^2c$

(E) $5a^7b^2c^{14}$

You can divide monomials like these by dividing their factors. Each term can be separated into factors involving a single variable or number. $10a^2b^8c^5$ has the factors 10, a^2, b^8, and c^5. Divide each of these by a corresponding factor in the denominator, $2a^5b^6c^6$.

$10 \div 2 = 5$

$a^2 \div a^5 = a^{2-5} = a^{-3}$

$b^8 \div b^6 = b^{8-6} = b^2$

$c^5 \div c^6 = c^{5-6} = c^{-1}$

The result of the division is the product of these four results: $5a^{-3}b^2c^{-1}$. Now we can rewrite this expression as a fraction. Any numbers and factors with positive exponents go in the numerator, and any with negative exponents go in the denominator (with the negative exponents changed to positive ones). So the numerator has $5b^2$, and the denominator has a^3c^1, or a^3c:

$$\frac{5b^2}{a^3c}$$

Choice (A) is the answer. Choice (B), $\frac{5ab^2c}{a^3c}$, includes the variables a and c in denominator, but variables raised to negative powers should appear in the denominator only. (D), $5a^3b^2c$, is the result of combining all of the factors with positive exponents in the numerator.

Question 4: Simplifying by Factoring

Which fraction is a simplified form of $\dfrac{2x^2 - 10x - 28}{x^2 - 10x + 21}$?

(A) $\dfrac{2x - 4}{x - 3}$

(B) $\dfrac{2x - 4}{x + 3}$

(C) $\dfrac{2x + 4}{x - 3}$

(D) $\dfrac{x + 4}{x - 3}$

(E) $\dfrac{x + 2}{x + 3}$

Like other fractions, this one can be simplified by dividing the numerator and the denominator by a common factor. To identify it, we need to factor each polynomial into binomials.

Since the first coefficient in $2x^2 - 10x - 28$ is 2, we are looking for factors $2x + j$ and $x + k$. The values j and k must be such that $j + 2k = -10$, and $jk = -28$. To start, list pairs of numbers that have a product of -28. We need to find the pair that satisfies $j + 2k = -10$.

Since the numbers have a negative product, we know that one is negative and the other is positive. Pairs with a product of -28:

$-28, 1$

$-14, 2$

$-7, 4$

$-4, 7$

$-2, 14$

$-1, 28$

Though it can be time-consuming, you need to plug each pair of numbers into $j + 2k = -10$, both ways, until you find a pair that makes the equation true.

If $j = 4$ and $k = -7$, then $j + 2k = 4 + 2(-7) = 4 - 14 = -10$. So the factors of $2x^2 - 10x - 28$ are $2x + 4$ and $x - 7$.

Since the first coefficient in $x^2 - 10x + 21$ is 1, factoring this polynomial should be more straightforward. We're looking for binomial factors with numbers that have a sum of -10 and a product of 21. -7 and -3 are those numbers, so $x^2 - 10x + 21 = (x - 7)(x - 3)$. So:

$$\frac{2x^2 - 10x - 28}{x^2 - 10x + 21} = \frac{(2x + 4)(x - 7)}{(x - 7)(x - 3)}$$

Since $x - 7$ appears in both the numerator and the denominator, divide top and bottom by $x - 7$.

$$\frac{2x + 4}{x - 3}$$

Choice (C) is the correct answer. Choice (A) is the result of factoring, $2x^2 - 10x - 28$ into $2x - 4$ and $x - 7$ instead of $2x + 4$ and $x - 7$.

Question 5: Solving Equations with Algebraic Fractions

If $\dfrac{10x^2 + 4x - 27}{3x + 3} = 3x - 1$, then $x =$

(A) -4 or 6

(B) -3 or 1

(C) -2 or 12

(D) 3 or -8

(E) 3 or -9

Here we must first get rid of the denominator in the fraction. We can accomplish that by multiplying both sides by $3x + 3$.

$10x^2 + 4x - 27 = (3x + 3)(3x - 1)$ Multiply binomials $3x + 3$ and $3x - 1$

$10x^2 + 4x - 27 = 9x^2 + 6x - 3$

We now must get this equation into the standard form of a quadratic equation, $ax^2 + bx + c = 0$. To do this, we can subtract $9x^2 + 6x - 3$ from both sides:

$10x^2 + 4x - 27 - (9x^2 + 6x - 3) =$

$(10x^2 - 9x^2) + (4x - 6x) + (-27 -(-3)) =$

$x^2 - 2x - 24$

$x^2 - 2x - 24 = 0$

Now we can factor this standard form quadratic equation. Following the steps we reviewed in chapter 11, we are looking for two numbers that have a product of c (-24) and a sum of $-b$ (2). The numbers -4 and 6 satisfy both of those conditions, so **(A) is the correct answer.**

Choice (B), -3 or -1, might be tempting because those numbers have a sum of b, -2. (C) gives the solutions to $x^2 + 10x - 24 = 0$. One might get that equation as a result of adding $6x$ to $10x^2 + 4x - 27$ instead of subtracting.

CHAPTER QUIZ

1. Which of these is undefined when $x = -4$?

   (A) $\dfrac{x+4}{x^2-4}$

   (B) $\dfrac{x-4}{x^2-4}$

   (C) $\dfrac{x+2}{4-x}$

   (D) $\dfrac{x+2}{x^2-16}$

   (E) $\dfrac{x-2}{x^2+16}$

2. $\dfrac{x+1}{x-4} + \dfrac{x-3}{x^2-6x+8} =$

   (A) $\dfrac{2x-4}{x^2-5x+4}$

   (B) $\dfrac{2x-2}{x^2-5x+4}$

   (C) $\dfrac{x^2-2x-5}{x^2-6x+8}$

   (D) $\dfrac{x^2-5}{x^2-6x+8}$

   (E) $\dfrac{x^2+2x-5}{x^2-6x+8}$

3. $\dfrac{2a+5}{a-6} - \dfrac{a+3}{a-2} =$

   (A) $\dfrac{6a+8}{a^2-8a+12}$

   (B) $\dfrac{a^2-2a-28}{a^2-8a+12}$

   (C) $\dfrac{a^2-2a+8}{a^2-8a+12}$

   (D) $\dfrac{a^2+4a-28}{a^2-8a+12}$

   (E) $\dfrac{a^2+4a+8}{a^2-8a+12}$

4. $\dfrac{4r^8s^2t^3}{12r^2s^4} =$

   (A) $\dfrac{r^4t^3}{3s^2}$

   (B) $\dfrac{r^6t^2}{3s^2}$

   (C) $\dfrac{r^6t^3}{3s^2}$

   (D) $\dfrac{r^6}{3s^2t^3}$

   (E) $\dfrac{r^6st^3}{3s^2}$

5. $\dfrac{a^2 - 12a + 27}{a - 3} =$

    (A) $a - 9$

    (B) $a + 15$

    (C) $a^2 - 9$

    (D) $a^2 + 4a - 9$

    (E) $a^2 - 4a + 9$

6. $\dfrac{x - 9}{2x^2 - 10x - 72} =$

    (A) $\dfrac{1}{2x - 8}$

    (B) $\dfrac{1}{2x + 8}$

    (C) $\dfrac{1}{2x + 16}$

    (D) $\dfrac{x}{2x + 8}$

    (E) $\dfrac{x}{2x + 16}$

7. If $\dfrac{a + 7}{a - 2} = 8$, $a =$

    (A) $-\dfrac{23}{9}$

    (B) $-\dfrac{9}{7}$

    (C) $\dfrac{7}{23}$

    (D) $\dfrac{9}{7}$

    (E) $\dfrac{23}{7}$

8. If $\dfrac{2y^2 + 8y - 24}{y + 4} = y + 6$, then $y =$

    (A) -6 or -8

    (B) -6 or 8

    (C) 0 or -2

    (D) 0 or 2

    (E) 6 or -8

ANSWER EXPLANATIONS

1. D

A fraction is undefined when the denominator has a value of 0. Since we are given the value of the variable, we can evaluate each denominator for $x = -4$:

A, B: $x^2 - 4 = (-4)^2 - 4 = 16 - 4 = 12$

C: $4 - x = 4 - (-4) = 4 + 4 = 8$

D: $x^2 - 16 = (-4)^2 - 16 = 16 - 16 = 0$

E: $x^2 + 16 = (-4)^2 + 16 = 16 + 16 = 32$

Since only $x^2 - 16 = 0$ when $x = -4$, $\dfrac{x+2}{x^2-16}$ is the only fraction that is undefined for that value.

Choice (A) might be tempting because the numerator has a value of 0 when $x = -4$, but a fraction with a numerator equal to 0 has that value and is not undefined.

2. D

To add fractions, you must have common denominators. But before multiplying these two denominators, check to see whether $x - 4$ is a factor of $x^2 - 6x + 8$. In fact, $x^2 - 6x + 8$ is the product of $x - 4$ and $x - 2$. So we can get fractions with common denominators by multiplying the top and bottom of $\dfrac{x+1}{x-4}$ by $x - 2$.

$$\dfrac{x^2 - x - 2}{x^2 - 6x + 8} \qquad \text{Combine the fractions}$$

$$\dfrac{x^2 - x - 2}{x^2 - 6x + 8} + \dfrac{x - 3}{x^2 - 6x + 8} = \dfrac{x^2 - 5}{x^2 - 6x + 8}$$

(B) is the result of adding the numerators and the denominators, without using fractions with common denominators as required.

3. E

Subtracting fractions follows the same rules as addition; we need common denominators. Here, the LCM of $a - 6$ and $a - 2$ is:

$$(a - 6) \times (a - 2) = a^2 - 2a - 6a + 12 = a^2 - 8a + 12$$

That will be the denominator of each rewritten fraction. Since we multiplied $a - 6$ by $a - 2$, the numerator of $\frac{2a+5}{a-6}$ must be multiplied by that binomial:

$$(2a + 5) \times (a - 2) = 2a^2 + 5a - 4a - 10 = 2a^2 + a - 10$$

$$\frac{2a+5}{a-6} = \frac{2a^2 + a - 10}{a^2 - 8a + 12}$$

Now multiply the numerator of $\frac{a+3}{a-2}$ by $a - 6$:

$$(a + 3) \times (a - 6) = a^2 + 3a - 6a - 18 = a^2 - 3a - 18.$$

$$\frac{a+3}{a-2} = \frac{a^2 - 3a - 18}{a^2 - 8a + 12}$$

$$\frac{2a+5}{a-6} - \frac{a+3}{a-2} = \frac{2a^2 + a - 10}{a^2 - 8a + 12} - \frac{a^2 - 3a - 18}{a^2 - 8a + 12} =$$

$$\frac{2a^2 + a - 10 - (a^2 - 3a - 18)}{a^2 - 8a + 12} =$$

$$\frac{2a^2 + a - 10 - a^2 + 3a + 18}{a^2 - 8a + 12} =$$

$$\frac{a^2 + 4a + 8}{a^2 - 8a + 12}$$

4.   **C**

Start by dividing the numbers 4 and 12. Since $4 \div 12 = \frac{1}{3}$, we can take the 4 out of the numerator and replace the 12 with 3.

$$\frac{4r^8 s^2 t^3}{12 r^2 s^4} = \frac{r^8 s^2 t^3}{3 r^2 s^4}$$ Divide the factors with the variable r. Since

$$r^8 \div r^2 = r^{8-2} = r^6 :$$

$$\frac{r^8 s^2 t^3}{3 r^2 s^4} = \frac{r^6 s^2 t^3}{3 s^4}$$ Since $s^2 \div s^4 = s^{2-4} = s^{-2} = \frac{1}{s^2}$:

$$\frac{r^6 s^2 t^3}{3 s^4} = \frac{r^6 t^3}{3 s^2}$$ Since t doesn't appear in the denominator, it cannot be divided

$$\frac{4 r^8 s^2 t^3}{12 r^2 s^4} = \frac{r^6 t^3}{3 s^2}$$

Choice (A) would result had you divided exponents instead of subtracted. (B) is a simplified form of $\frac{4 r^8 s^2 t^3}{12 r^2 s^4 t}$.

5. A

We can try to simplify this algebraic fraction by finding a common factor in the numerator and the denominator. The first step is to factor the numerator, $a^2 - 12a + 27$ into binomials. That trinomial is the product of $a - 9$ and $a - 3$. So the numerator and the denominator have a common denominator of $a - 3$.

Dividing both the top and the bottom of the fraction leaves us with $\frac{a-9}{1}$ or $a - 9$. Choice (C), $a^2 - 9$, is what one would get by finding $a^2 - 9$ instead of $a - 9$ as a factor of $a^2 - 12a + 27$.

6. B

The first step is to find a common factor in $x - 9$ and $2x^2 - 10x - 72$. Since $x - 9$ cannot be factored in polynomials, look at $2x^2 - 10x - 72$. That trinomial is the product of $2x + 8$ and $x - 9$. So $\frac{x-9}{2x^2 - 10x - 72} = \frac{x-9}{(2x+8)(x-9)}$, and $x - 9$ is a factor of both polynomials.

Dividing both the numerator and the denominator by that binomial, we get $\frac{1}{2x+8}$, which is the fully simplified form of the fraction.

7. E

Multiply both sides of the equation by $a - 2$. That binomial cancels out on the left side, and is multiplied by 8 on the right side.

$a + 7 = 8(a - 2) = 8a - 16$

So we must solve $a + 7 = 8a - 16$. Subtracting a from both sides and then adding 16 gets us $23 = 7a$, or $7a = 23$. The last step, dividing both sides by 7, gets us $a = \frac{23}{7}$.

8. B

First multiply both sides of $\dfrac{2y^2 + 8y - 24}{y + 4}$ by $y + 6$ to get rid of the denominator:

$2y^2 + 8y - 24 = y^2 + 10y + 24$ Subtract the term in $y^2 + 10y + 24$ from both sides

$y^2 - 2y - 48 = 0$

$y^2 - 2y + 48 = (y + 6) \times (y - 8)$

$y = -6$ or $y = 8$

Word Problems

WHAT ARE WORD PROBLEMS?

Word problems are math questions presented without the symbols of algebra and arithmetic. Many Algebra I concepts can be applied in the form of word problems to real-world situations. The challenge lies in writing the word problem in such as way that it preserves the symbolic statements of algebra. The key task is to identify the unknown quantity in the given scenario and find its place in the right equation or inequality.

Algebraic word problems cover many subjects, many of which can be approached with key formulas that can be memorized. Some problems, of course, do not involve common formulas, but can still be set up in terms of equations or inequalities.

CONCEPTS TO HELP YOU

A formula represents an important relationship between numbers and symbols. Applying a formula to a word problem is a matter of plugging in the values you are provided with in the question, and then solving for whatever variable is still unknown.

Percentages

Many algebra word problems involve percentages. Percentages come up in many real–world situations, such as taxes and interest on loans. A percentage is a special way of expressing a fraction or decimal. One percent—or 1%—is equal to $\frac{1}{100}$ or 0.01. 100% is equal to one.

The value of any percentage, in fraction or decimal form, is the number divided by 100:

$$57\% = \frac{57}{100} \text{ or } 0.57$$

To find a certain percent of a number, multiply it by the corresponding decimal or fraction

$$x\% \text{ of } y = y \times \frac{x}{100}$$

Area and Perimeter

Area and perimeter are important concepts in Geometry as well as Algebra. Here, we'll focus on two key geometric figures: rectangles and circles.

The *area* of a rectangle is the amount of space inside of it. It is the product of the rectangle's length and width. The area formula, then, is:

$$a = lw$$

where a is the area of the rectangle, l is its length, and w is its width.

The perimeter of a rectangle is the sum of its side lengths. Since two sides share the dimension of a rectangle's length, and the other two sides share the dimension of its width, the perimeter formula is:

$$p = l + l + w + w \text{ or } p = 2l + 2w$$

where p is the perimeter.

The area of a circle depends on its radius, which is the distance from a point on the circle to its center. The area of any circle is equal to the square of its radius multiplied by a special number we call *pi*. Pi is often represented by the Greek letter π. Its value is roughly 3.14, but it is in fact a non–terminating, non–repeating decimal—an irrational number. So the area formula for the circle is $a = \pi r^2$.

The perimeter of the circle is the length of its perimeter. It is often called the *circumference*. The circumference of a circle is given by the formula $c = 2\pi r$.

Averages

The average of a set of numbers, *a*, is the sum of the numbers, *s*, divided by the number of members, *n*, in the set:

$$a = \frac{s}{n}$$

In Algebra I, you might be given the value of the average and then asked to find the value of the number of members, or of one of the numbers, based on other information provided.

Rates

A rate is basically a comparison of two measurements. Usually, one of the measurements is time. Velocity, for example, is a rate, as it compares a distance traveled to time elapsed. That is why velocity is described in terms of units such as *miles per hour* or *feet per second*. Another rate might be that of the speed of a photocopier or printer, described in terms of copies per second or pages per minute.

The standard rate formula is:

rate × time = distance [$rt = d$]

It is also common to determine rate in terms of the distance and time:

$$r = \frac{d}{t}$$

Many questions involving rates do not involve distances, strictly speaking, but you can think of distance of the total done in a given period. If we were talking about a photocopier, for instance, the "distance" might be the number of copies made in given period.

Inequalities

Algebraic word problems can involve inequalities as well as equations. Whereas some values can be described as equal, others are *greater than* or *less than* a given value.

"At least" probably means you're dealing with an inequality involving a *greater than or equal to* relationship

"At most" probably means you're dealing with an inequality involving a *less than or equal to* relationship

STEPS YOU NEED TO REMEMBER

The key to solving a word problem is knowing how to restate the described relationship in algebraic terms. This is a matter of identifying the equation or inequality that you need to work out in the question—the known and unknown values—and working those values into the statement.

Identifying the Relationship

When the question involves a relationship you're familiar with, such as one involving rate, time, and distance, you should be ready to apply the appropriate formula. If the question describes a new situation, where none of the formulas we reviewed earlier apply, you need to try restating the information in terms of a formula. What operations are described? What does the question say about the order in which they should be performed? As you answer those questions, you'll be in a better position to state the equation or inequality presented in the question.

Using the Known and Unknown Values

Once you have identified the equation or inequality that captures the described relationship, you need to identify the variables that have stated values. You can plug those into the statement, so as to have a statement with fewer variables and more numbers. Those variables with no stated values are the unknowns; they're the ones you must solve for.

When writing equations, use letters that will help you keep track of variables. There's a reason why we use the letters r, t, and d in the rate formula; after all, we'd be more likely to plug the value of the rate in the wrong place if we used letters such as x, y, and z.

STEP–BY–STEP ILLUSTRATION OF THE 5 MOST COMMON QUESTION TYPES

Now it's time to work through some common word problem types. They can cover a broad range of Algebra I topics, such as linear equations, quadratic equations, systems of equations, and inequalities.

Question 1: Percentage

Roger paid $2.80 in sales tax for a coat he purchased. If the sales tax rate is 4%, what was the sale price of the coat?

(A) $67.20

(B) $70.00

(C) $112.00

(D) $268.8

(E) $291.20

The unknown in this question is the sale price, of which $2.80 is a percentage share. Since $2.80 is 4% of the unknown variable x, then $2.80 is the product of x and 0.04. So $0.04x = 2.8$.

$$0.04x = 2.8 \qquad \text{Divide both sides by 0.04}$$
$$2.8 \div 0.04 = 70$$
$$x = 70$$

So the price of the coat is $70.00, and **(B) is the answer.** (A), $67.20, is just $2.80 subtracted from $67.20. When solving an algebraic word problem, take care to work all of the information provided into a mathematical relationship. Choice (D), $268.80, is the result of multiplying 2.8 by 96 (the difference between 100 and 4).

Question 2: Length and Area

The length of a rectangular room is 5 feet more than its width. If the area of the room is 204 square feet, what is the length in feet of the room?

In this area question, the length and width are both unknown. But we don't need two variables to relate the length and width to the area. That is because we can express the length of the room in terms of the width. Since the length is 5 feet more than the width, the variable x can stand for the width, and $x + 5$ can be the value of the width. The area is the product of the length and the width, so:

$x \bullet (x + 5) = 204$ Simplify left side

$x^2 + 5x = 204$ We see now that we're dealing with a quadratic equation. Get it into the standard form $ax^2 + bx + c = 0$. Subtract 204 from both sides

$x^2 + 5x - 204 = 0$ Factor c, -204, into two numbers that have a sum of $-b$ (which is -5)

Since $-17 \bullet 12 = -204$, and $-17 + 12 = -5$, the numbers -17 and 12 are the two solutions of $x^2 + 5x - 204 = 0$. One of these values of x must be the value of the room's width. Since a width cannot be negative, the width must be 12. The length is 5 feet more than this, and so **17 is the answer.**

Check this by multiplying the dimensions of the room. 12 and 17 have a product of 204, which was given as the area. Since 12 is the other dimension of the room, that's tempting, but incorrect.

Question 3: Averages

The average height of four men, Ben, Josh, Paul, and Sam, is 68 inches. Ben is 70 inches tall, Josh is 74 inches tall, and Paul is 66 inches tall. Sam's height is

(A) 52 inches

(B) 62 inches

(C) 69.5 inches

(D) 70 inches

(E) 72 inches

The average of a set of values is their sum divided by the number of values. Here, we're given the average (68); the number of values (4); and three of the four values (70, 74, and 66). The unknown is the fourth value, Sam's height. So the average height is given by the equation:

$$a = \frac{b + j + p + s}{n}$$

where a is the average, n is the number of values, b is Ben's height, j is Josh's height, p is Paul's height, and s is Sam's height. Since $a = 68$, $n = 4$, $b = 70$, $j = 74$, and $p = 66$:

$$68 = \frac{70 + 74 + 66 + s}{4}$$

$$68 = \frac{210 + s}{4} \qquad \text{Solve for } s \text{ by multiplying both sides by 4}$$

$$272 = 210 + s \qquad \text{Subtract 210 from both sides}$$

$$s = 62$$

Sam is 62 inches tall, so **(B) is correct.** (A), 52 inches, would result had you switched the values of Paul's height and the average. (C), 69.5, is actually just the average of the four values given.

Question 4: Inequalities

Kyle drove continuously for 4 hours, during which time he traveled between 196 and 224 miles. The possible range of his average speed, where r is the speed in average miles per hour, could be expressed with which of the following inequalities?

(A) $r \geq 28$

(B) $r \geq 52.5$

(C) $r \leq 52.5$

(D) $48 \leq r \leq 56$

(E) $r \leq 48$ or $r \geq 56$

This question requires us to apply the rate formula to solve a compound inequality. Specifically, we're asked to solve for the rate, which is the unknown value here. We know that the distance Kyle traveled was greater than 196 miles but less than 224 miles.

The rate formula is $rt = d$, where r is the rate, t is the time, and d is the distance. Since $d > 196$ and $d < 224$:

$rt > 196$ and $rt < 224$

Since $t = 4$, we can plug that value into each inequality:

$4r > 196$ and $4r < 224$

Solving each inequality for r, then, is a matter of dividing both sides of each by 4:

$r > 48$ and $r < 56$

Since r is greater than or equal to 48 and less than or equal to 56:

$48 < r < 56$

Choice (D) is the correct answer. (B) and (C), $r > 52.5$ and $r < 52.5$, are based on the average of 196 and 224 divided by 4.

Question 5: Systems of Equations

A movie theater charges different prices to adults and children. A group with 3 adults and 4 children bought tickets for a movie, spending a total of $40.50. A group with 1 adult and 3 children also bought tickets, spending a total of $21.00. What is the price of an adult ticket?

(A) $4.50

(B) $5.50

(C) $6.75

(D) $6.90

(E) $7.50

This question involves a system of equations, which we covered in chapter 8. Each of the total expenses, $40.50 and $21.00, is a sum of a combination of variables. In fact, there are two unknowns to deal with (though we're asked for only one): the price of a child's ticket and the price of an adult ticket.

Let's use a to stand for the adult ticket price, and c for the child ticket price. Since 3 adult tickets and 4 child tickets costs $40.50:

$$3a + 4c = 40.5$$

and since 1 adult ticket and 3 children's tickets costs $21.00:

$$a + 3c = 21$$

Since we have two equations, we can solve for both variables. We do this by combining equations or by substitution, but since combination would involve solving for both variables, substitution may be the better choice. It will be easier to solve the second equation for a.

$a + 3c = 21$	Subtract $3c$ from both sides
$a = 21 - 3c$	Substitute the value of a, $21 - 3c$, for a in the other equation
$3a + 4c = 40.5$	
$3(21 - 3c) + 4c = 40.5$	
$63 - 9c + 4c = 40.5$	
$63 - 5c = 40.5$	Subtract 63 from both sides
$-5c = -22.5$	Divide both sides by -5
$c = 4.50$	

So a child's ticket is $4.50. To find the price of an adult ticket, we plug that value into one of the equations, and then solve for a. Since $c = 4.50$, $a = 21 - 3(4.50) = 21 - 13.50 = 7.5$. So an adult ticket is $7.50, and **Choice (E) is correct.**

(B), $5.50, is what you would get had you substituted 4.5 for a instead of c in $a + 3c = 21$, and then solved for the other variable.

CHAPTER QUIZ

1. Janice correctly answered 62.5% of the questions on a test. If she answered 40 questions correctly, how many questions were on the test?

   (A) 62

   (B) 64

   (C) 65

   (D) 67

   (E) 70

2. A rectangular garden has an area of 3,268 square feet. If the length of the garden is 76 feet, how many feet wide is the garden?

3. The circular table top has an area of 706.5 square inches. What is the length of the radius of the table? (Use 3.14 for pi)

   (A) 15 inches

   (B) 25 inches

   (C) 30 inches

   (D) 112.5 inches

   (E) 225 inches

4. In a recent basketball game, one team scored 84 points. If the average number of points scored by a team member was 12 points, how many team members were there?

5. A bicycle shop rents bicycles for a charge of $8, plus $4 per hour. A rental for how many hours would cost $36?

   (A) 4

   (B) 5

   (C) 7

   (D) 9

   (E) 11

6. Ms. Thomson gave at least two pencils to each student in her class for a test. She had 22 pens left over. Which inequality indicates the number of pens Ms. Johnson started with, where s is the number of students in the class and p is the number of pens?

   (A) $p \le \dfrac{s}{2} - 22$

   (B) $p \le \dfrac{s}{2} - 11$

   (C) $p \le \dfrac{s}{2} + 22$

   (D) $p \ge 2s - 22$

   (E) $p \ge 2s + 22$

7. A pot that can hold 10 liters already contains 1.5 liters of water. If the pot is then filled from a faucet at a rate of at least 0.1 liters per second, which inequality states how long it will take to fill the pot, where t is the number of seconds?

(A) $t \leq 0.85$

(B) $t \geq 1.15$

(C) $t \leq 85$

(D) $t \geq 85$

(E) $t \leq 115$

8. A supermarket sells cases of canned fruit. A large size case holds more cans than the regular size cases. Together, five large cases and three regular cases hold 240 cans. And together, three large cases and two regular cases hold 148 cans. How many cans does one large case hold?

(A) 18

(B) 20

(C) 24

(D) 30

(E) 36

ANSWER EXPLANATIONS

1. B

If 40 is 62.5% of the unknown total number of questions, then $0.625x = 40$, when x is that total number. To get the variable alone, divide both sides by 0.625. Since $40 \div 0.625 = 64$, that is the total number of questions.

Choice (C) is the product of 40 and 1.625. This error might result had you taken 1.625 to be the multiplicative inverse of 0.625. Since 0.625 is $\frac{5}{8}$, the multiplicative inverse is $\frac{8}{5}$, or 1.6.

2. 43

To answer this area question, we'll need to plug the known values into the area formula. That will allow us to solve for the unknown—the width. The area formula is $a = lw$, where a is the area and l is the length.

$3268 = 76w$ Divide both sides by 76

$3268 \div 76 = 43$

So 43 feet is the width of the table. One common incorrect answer to this question is 1,558, which is the width if 3,268 were the *perimeter* of the yard rather than the area. Be sure that the formula you use applies to the right property raised in the question.

3.   A

The formula for the area of a circle is $a = \pi r^2$, where r is the radius. Here we're given the area of the circle, and the unknown value is the radius. To find that value, we can plug in the area, as well as the rounded value of π the question tells us to use.

$$3.14r^2 = 706.5 \qquad \text{Divide both sides by 3.14}$$
$$r^2 = 225 \qquad \text{Since } r \text{ is the square root of } r^2, \text{ we see that}$$
$$r = \sqrt{225} = 15$$

So the radius of the table is 15 inches. You would get choice (B) if you got 25 instead of 15 as the positive square root of 225. (C), 30 inches, is actually the correct diameter of the circle.

4.   7

The average of a set of values is their sum divided by the number of values. We can represent this relationship with the formula:

$$a = \frac{s}{n}$$

where a is the average, s is the sum of the values, and n is the number of values.

$$12 = \frac{84}{n} \qquad \text{Multiply both sides by } n$$
$$12n = 84 \qquad \text{Divide both sides by 12}$$
$$n = 7$$

5.   C

If h is the number of hours of the rental, and the cost of renting a bicycle in addition to the $8 flat fee is $4 per hour. the total rental cost is $4h + 8$. Since the total charge for the rental in question is $36:

$$4h + 8 = 36 \qquad \text{Subtract 8 from both sides}$$
$$4h = 28 \qquad \text{Divide both sides by 4}$$
$$h = 7$$

So a rental of 7 hours costs \$36.

6. E

Every student got at least two pens, and there are an additional 22 pens. What if every student got exactly two pens? Then p would equal $2s + 22$. For instance, if there were 20 students, then 40 pens would be given out, and the remaining 22 would bring the total to 62.

Now since each student got *at least* two pens, the number could be even higher. That means that p is greater than or equal to the sum of $2s$ and 22.

7. C

Since the faucet can fill the pot at a rate of at least 0.1 liters per second:

$$r \geq 0.1$$

where r is the rate. Now, the volume of water filled can be given with the formula:

$$rt = v$$

where v is the volume. This is a variation on the standard rate formula, $rt = d$. If $rt = v$, then $r = \dfrac{v}{t}$. The value of v is 8.5, as that is the space remaining in the 10 liter pot, given that it already contained 1.5 liters of water. Since $r \geq 0.1$:

$$\frac{v}{t} \geq 0.1$$

$$\frac{8.5}{t} \geq 0.1 \qquad \text{Multiply both sides by } t$$

$$8.5 \geq 0.1t \qquad \text{Divide both sides by 0.1}$$

$$85 \geq t$$

$$t \leq 85$$

Choice (D), $t \geq 85$, could be the result of changing the direction of the inequality sign when dividing. The sign direction changes only when dividing or multiplying by a negative number.

8.   E

Here we have two unknowns: the number of cans each case holds and two algebraic relationships. Thus, we need to treat this as a system of equations. Let's use l to represent the number of cans in a large case and r to represent the number of cans in a regular case. With those variables, we can write out two equations:

$$5l + 3r = 240$$
$$3l + 2r = 148$$

Let's solve the first equation for r

$5l + 3r = 240$	Subtract $5l$ from both sides
$3r = 240 - 5l$	Divide both sides by 3
$r = 80 - \dfrac{5l}{3}$	Substitute this value for r into $3l + 2r = 148$
$3l + 2(80 - \dfrac{5l}{3}) = 148$	Rewrite
$3l + 160 - \dfrac{10l}{3} = 148$	Combine the variables and subtract 160 from both sides
$-\dfrac{l}{3} = -12$	Multiply both sides by $-\dfrac{1}{3}$
$l = 36$	

So the large case holds 36 cans. Choice (A), 18, is the result of getting $-\frac{2l}{3}$ instead of $-\frac{l}{3}$ when subtracting $\frac{10l}{3}$ from $3l$. Choice (B), 20, is actually the number of cans a regular size case holds.